Dairy Animal Production: Scientific and Sustainable Methods

Dairy Animal Production: Scientific and Sustainable Methods

Editor

Carin Fisker

scitus academics

Dairy Animal Production: Scientific and Sustainable Methods

Edited by **Carin Fisker**

ISBN: 978-1-68117-421-1
Library of Congress Control Number: 2016937429

SCITUS Academics LLC,
www.scitusacademics.com
Box No. 4766, 616 Corporate Way,
Suite 2, Valley Cottage,
NY 10989

Notice

Reasonable efforts have been made to publish reliable data and views articulated in the chapters are those of the individual contributors, and not necessarily those of the editors or publishers. Editors or publishers are not responsible for the accuracy of the information in the published chapters or consequences of their use. The publisher believes no responsibility for any damage or grievance to the persons or property arising out of the use of any materials, instructions, methods or thoughts in the book. The editors and the publisher have attempted to trace the copyright holders of all material reproduced in this publication and apologize to copyright holders if permission has not been obtained. If any copyright holder has not been acknowledged, please write to us so we may rectify.

Printed in United States of America on Acid Free Paper

Preface

World milk production is almost entirely derived from cattle, buffaloes, goats, sheep and camels. Other less common milk animals are yaks, horses, reindeers and donkeys. The presence and importance of each species varies significantly among regions and countries. The key elements that determine the dairy species kept are feed, water and climate. Other factors that may influence the presence of a dairy species are market demand, dietary traditions and the socio-economic characteristics of individual households. Although cattle are kept in a wide range of environments, other dairy species make dairying possible in adverse environments that often cannot support any other type of agricultural production. Sheep allow milk production in semi-arid regions around the Mediterranean, goats in regions with poor soils in Africa, horses in the steppes of Central Asia, camels in arid lands, buffaloes in wet tropical regions, and yaks in high mountainous areas such as the Tibetan Plateau. In developing countries, milk producing animals are often raised in subsistence and smallholder systems. These animals are usually multi-purpose and grow and produce under difficult conditions, such as low inputs, minimum management and harsh environments. They are well adapted to local conditions, but have low genetic potential for milk production. Dairy cattle are cattle cows bred for the ability to produce large quantities of milk, from which dairy products are made. Dairy cows generally are of the species Bos taurus. Historically, there was little distinction between dairy cattle and beef cattle, with the same stock often being used for both meat and milk production. Today, the bovine industry is more specialized and most dairy cattle have been bred to produce large volumes of milk. Dairy cows may be found either in herds on dairy farms where dairy farmers own, manage, care for, and collect milk from them, or on commercial farms. Herd sizes vary around the world depending on landholding culture and social structure. The book Dairy Animal Production: Scientific and Sustainable Methods deals with basic management aspects and production methods of dairy animal. The informations are carefully presented in concise manner and to the point.

Table of Contents

CHAPTER 1

Nutrition and Health of Dairy Animals

Cazim Crnkic[1] and Aida Hodzic[1]

[1] University of Sarajevo, Veterinary faculty, Bosnia & Herzegovina

1. INTRODUCTION

Modern breeds of dairy animals are able to produce huge amount of milk. In attempt to consume, digest and metabolize enough nutrients to satisfy lactation needs, those animals are exposed to serious stress conditions that can affect their health. Health problems which arise from those conditions are mainly related to impaired ability to metabolize enough nutrients to compensate for those lost in milk. They are known as metabolic or production diseases and may be of great economic importance in milk production systems.

Although metabolic diseases have become a common problem on dairy farms, they still require a serious attention to be controlled. The incidences of these disorders can be reduced by proper nutrition of animals. Also, some of the specific strategies in feeding practice offer additional advantages in prevention of nutrition-related metabolic diseases.

2. NUTRITION-RELATED METABOLIC DISORDERS, THEIR ETIOLOGY AND CONSEQUENCES

The most important metabolic diseases in dairy cows, ewes and goats can be discussed as energy and/or fiber related, lipid related or vitamin and mineral related disorders. However, they are not easily categorized according to their cause, thus the pathogenesis and hallmarks of each disease should be considered for their categorization. Energy related disorders (related to energy density of the diet or feed intake) include fatty liver and ketosis, rumen acidosis, laminitis, displaced abomasums and milk fat depression. As energy density and fiber content of the diets are often inversely related, most of these diseases can be considered as fiber related too. Fatty liver and ketosis also can be categorized as lipid related disorders due to changes in lipid metabolism in affected animals (Grummer, 1993). The most important mineral/vitamin related disorders are hypocalcaemia, hypomagnesaemia, udder edema, retained placenta and metritis. Not all of them are ultimately caused by mineral composition of the diet, but may be prevented by manipulation of minerals or vitamins in the diet. Broken homeostatic mechanisms often become major etiological factors in the development of metabolic disease, and consequences varies from reduced production and impaired reproductive performances to increased risk to develop other diseases.

2.1. Fatty Liver

Fatty liver (hepatic lipidosis, fat cow syndrome) is a metabolic disorder characterized by a high content of lipids and triglycerides in the liver (Ingvartsen, 2006). The disease occurs in periparturient period, primarily in the first 4 weeks after calving (Grummer, 1993), and as a secondary disease of other production diseases that depress appetite or increase body fat mobilization. The clinical symptoms comprise depression, lack of appetite and weight loss, and the cows are weak and apathetic (Radostits et al., 2000). Most cows suffer from non-specific clinical signs including reduced rumen motility and decreased milk yield. However, this disease occurs especially in its subclinical form and can be a problem for up to 50% of cows in early lactation (Ingvartsen, 2006).

Risk factors for fatty liver in dairy cows may be nutritional, managerial, and genetic. Prepartum risk factors include obesity, severe feed restriction, feeding excess energy, and long calving interval. Postpartum risk factors are different diseases and infection, fasting and feed restriction, as well as ketogenic diets and sudden feed changes (Bobe et al., 2004). Among the highest risk factors of fatty liver is a high rate of mobilization of body lipids around calving in overconditioned cows which have been overfed in late lactation and the dry period and had low feed intake around calving (Ingvartsen, 2006). Increased prepartum body weight gain or body condition increase liver fat content and indicate increased risk of fatty liver and related disorders. Therefore, the primary nutritional risk factor for fatty liver is obesity. In obese cows, lipolysis of adipose tissue is increased more during peripartal period than in cows with normal body condition. Obese cows have a greater decrease in feed intake around calving and, therefore, have a more severe negative energy balance (Bobe et al., 2004). During the negative energy balance body fat is mobilized into the bloodstream in the form of non-esterified fatty acids (NEFA). NEFA are taken by the liver in proportion to their supply, but the liver does not have capacity sufficient to oxidize and use all amount of NEFA for energy. Therefore, cows are predisposed to accumulate NEFA as triglycerides within the liver when large amounts of NEFA are released from adipose tissue (Overton & Waldron, 2004). Rate of triglyceride synthesis is proportional to plasma NEFA concentration, thus fatty liver is likely to develop when plasma NEFA are elevated (Grummer, 1993).

Severe negative energy balance caused by low feed intake around calving leads to increase in body fat mobilization and plasma NEFA concentration. Thus, the severity of fatty liver may be reduced if drop in feed intake before calving is prevented (Bertics et al., 1992).

Diets deficient in cobalt (Co) have been shown to cause fatty liver in sheep. It has been termed as ovine white liver disease. Hepatic lipidosis in goats due to feeding low levels of Co in their diet has also been reported (Johnson et al., 2004).

Fatty liver is associated with decreased health status and reproductive performance. In severe cases, milk production and feed intake are also decreased (Bobe et al., 2004). Although fatty liver is often a reversible condition, it predisposes cows to reduced liver function and to a number of other production diseases (Ingvartsen, 2006). The incidence of fatty liver is strongly associated with the incidence of especially ketosis and displaced abomasum because these disorders are related to severe negative energy balance too (Bobe et al., 2004). Inhibition of gluconeogenesis may also occur when triglycerides accumulate in liver (Overton & Waldron, 2004) and additionally increase the risk for ketosis. Different aspects of the immune response are also suppressed in cows with fatty liver (Bobe et al., 2004; Overton & Waldron, 2004) and those animals are more prone to infectious diseases. However, the exact costs of fatty liver are difficult to estimate because condition can be diagnosed only by liver biopsy (Bobe et al., 2004).

2.2. Ketosis

Ketosis (acetonemia) is a metabolic disorder characterized by elevated concentrations of the ketone bodies (acetoacetate, beta-hydroxybutyrate and acetone), and a low to normal concentration of glucose in the blood. This disorder occurs both subclinically and clinically. Prevalence of subclinical ketosis is the highest in the first 2 weeks of lactation ranging from 8,9% to 34% in various studies (Ingvartsen, 2006; Rushen et al., 2008). Blood glucose in clinically affected cows fall below the level required to support nerve and brain function and cows often exhibit signs of central nervous system dysfunction. Ketotic cows also suffer inappetence which further exacerbates the negative energy balance. Milk production falls seriously which helps the cow to cope with the negative energy balance (Goff, 2006a).

Four main types of ketosis are: primary ketosis, secondary ketosis, butyric acid ketosis and underfeeding ketosis (Ingvartsen, 2006). Classical or primary ketosis (also called spontaneous or production ketosis) generally occurs in cows during the first 2 to 4 weeks of lactation (Goff, 2006a). It develops when the glucose demand exceeds the gluconeogenesis capacity of the liver resulting in increased ketogenesis, and thus in high concentrations of ketone bodies in blood, milk, and urine. The disorder is mainly seen in obese cows. Secondary ketosis results from another disease that depresses feed intake and increases body fat mobilization (Ingvartsen, 2006).

Butyric acid ketosis is caused by large amounts of butyrate in the feed and probably by related depressed feed intake. Silage with high butyrate concentrations results in an increased concentration of beta-hydroxybutyrate in the blood, and such silage is consumed in lower amount then normal one. Underfeeding ketosis occurs especially in cows that are fed insufficiently. Underfed animals are deficient in glucogenic precursors and this condition then leads to increased ketogenesis (Ingvartsen, 2006).

Another form of ketosis often seen in US dairies usually occurs in cows less than 10 days in lactation. It can often be difficult to treat as is generally accompanied by some degree of fatty liver (ketosis/fatty liver complex). Ketosis, like fatty liver, also occurs during periods of elevated plasma NEFA. Clinical symptoms are similar to classical ketosis, but ketotic state was actually preceded by an increase in fat accumulation in the liver (Goff, 2006a).

Major factors directly or indirectly increasing the risk of fatty liver and ketosis are: overconditioning at calving, excessive mobilization of body fat, low nutrient intake, some nutrient or diet specific factors and management and environmental stress (Ingvartsen, 2006). Ketosis is a metabolic condition that occurs primarily when a cow is in negative energy balance immediately after calving. To support the energy demands, the body mobilizes fat reserves. Excessive release of NEFA from fat depots may overwhelm the capacity of the liver to use the fatty acids as a fuel. They are instead converted to ketone bodies (Rushen et al., 2008). Feed intake is naturally depressed by about 20% around the time of calving. Simultaneous with the feed intake depression, plasma NEFA and liver triglyceride concentrations exhibited their greatest increase (Goff, 2006a). All factors exacerbating negative energy balance and depression of feed intake around calving increase the risk of ketosis, especially in overconditioned cows.

Negative energy balance, high blood levels of NEFA or beta-hydroxybutirate and ketosis reduce the response capacity of white blood cells so that invading bacteria can out compete the immune system of the cow (Zadoks, 2006). Hypoglycemia alone is not likely to exacerbate periparturient immunosuppression, but hyperketonemia appears to have multiple negative effects on immune functions (Overton & Waldron, 2004). Impaired immune functions around calving make cow more prone to infectious diseases. Cows that have suffered ketosis have a higher risk of ketosis in the following lactation (Ingvartsen, 2006). Affected animals produce 0,7-3,3 kg/day less milk in the rest of lactation (Fourichon et al., 1999) or about 200 kg annually (Guard, 1996). In addition, prolonged and pronounced negative energy balance in early lactation adversely affects reproductive performance of cows and increases the risk of other metabolic diseases (Rushen et al., 2008).

Ewes experience ketosis typically during the last month of pregnancy and is known as pregnancy toxemia (Schlumbohm & Harmeyer, 2003). Ovine pregnancy toxemia (twin lamb disease, pregnancy disease) occurs primarily in ewes carrying more than one fetus (Andrews et al., 1996; Schlumbohm & Harmeyer, 2003). It is usually seen when the plane of nutrition has been static or falling over the last month of gestation when the requirements of the ewe are increasing to allow the growth of the fetuses (Andrews et al., 1996). In goats, two clinical form of ketosis have been described: pregnancy toxemia during the last month of pregnancy and primary ketosis during the first month of lactation (Stelletta et al., 2008).

Limitation of the availability of glucose has been found to be a crucial factor for the development of pregnancy toxemia. Hyperketonemia lowered plasma glucose concentration and depressed endogenous glucose production in sheep by approximately 30%, and this facilitates the onset of pregnancy toxemia (Schlumbohm & Harmeyer, 2003).

Mortality is high if pregnancy toxemia is not treated, with only about 20% of affected ewes recovering without treatment. Encephalopathy can results from depressed glucose metabolism in the brain (Andrews et al., 1996).

2.3. Rumen Acidosis

Ruminal acidosis is a nutritional disorder of ruminants generally resulting from ingestion of large amounts of feeds rich in readily fermentable carbohydrates, particularly when animals have not been gradually acclimated to those feeds (Bramley et al., 2008). The disorder usually occurs in high-producing dairy cows. Subacute or subclinical ruminal acidosis (SARA) is considered to be one of the major threats to the welfare of lactating dairy cows and may affect up to 20% of cattle in early to mid lactation (Rushen et al., 2008). Subclinical rumen acidosis is defined as a condition where rumen fluid pH is below 6,0 while acute rumen acidosis is when rumen pH is below 5,5 associated with rumen motility that is weak or ceased (Ingvartsen, 2006).

Rumen acidosis classically occurs when an animal consumes the excess of grain (Stone, 2004). Fermentation of the high grain diet reduces rumen pH which can cause undesirable changes in microbial populations within the rumen (Goff, 2006a). Rumen pH is lowered due to large quantities of volatile fatty acids and lactic acid produced during grain fermentation (Bramley et al., 2008). Lactic acid production is not the hallmark of rumen acidosis in dairy cows, as observed in beef feedlot cattle. Instead, it is the total organic acid load induced by the high grain diet combined with the inability of the cow to buffer the acids with salivary secretions (Goff, 2006a; Stone, 2004). However, lactate accumulation may occur in cows after calving if the shift in fermentable carbohydrates between the diets fed before and after calving is too dramatic (Stone, 2004).

Low rumen pH can cause rumenitis, metabolic acidosis, lameness, hepatic abscesses formation, pneumonia, and even death (Bramley et al., 2008). Reduced ruminal efficiency, liver and lung abscesses, and laminitis all can be related to SARA (Stone, 2004).

2.4. Laminitis

Laminitis and laminitis-related hoof problems (sole ulcer, white line abscess, solar hemorrhage, etc.) are one of the leading causes of lameness in cows. Laminitis has been associated with nutrition, and specifically with ruminal acidosis either in its acute or subacute form. Exact relationship between laminitis and SARA is not known. One of the theories states that damage of the ruminal epithelium induced by acidosis allows absorption of histamine and endotoxins into the blood. These and possibly other compounds affect circulation within the hoof and cause inflammation leading to the condition known as laminitis (Stone, 2004). Cows fed the higher level of crude protein may have increased incidence and duration of lameness. It is considered that products of degradation of protein excess in the rumen may be the causative agents for lameness.

It is widely accepted that nutrition helps maintain claw health through the production of good quality horn (Baird & Muelling, 2009). For example, heifers fed on hay before calving might have better foot health than those fed silage, even if both groups are being fed the same diets postpartum. Forage type and level of moisture in the diet could influence lameness and this effect is exerted even before first calving (Logue & Offer, 2001).

It has also been shown that nutritional supplements such as biotin and zinc (Zn) can help reduce lameness through improving claw horn quality (Baird & Muelling, 2009; Goff, 2006a). Biotin is essential for two major metabolic pathways in keratinisation, keratin protein synthesis and lipogenesis. Other vitamins also may play important roles in maintaining claw integrity, including vitamin A and vitamin E. Zinc, as a component of many enzyme systems, has a role in major functions during keratinisation including the formation of structural proteins (Baird & Muelling, 2009). This element improves claw integrity by speeding wound healing and epithelial tissue repair. Other trace minerals that impact claw condition include iodine (I), selenium (Se), copper (Cu), manganese (Mn) and Co. Calcium and phosphorus (P) are also needed for normal claw growth and integrity.

Laminitis reduces profitability of the dairy herd. It is estimated that 15% of cows culled for slaughter are culled due to laminitis. In clinically lame cows, milk yield was reduced from 4 months before and for the 5 months after treatment. The total mean estimated reduction in milk yield per lactation was approximately 360 kg (Green et al., 2002).

2.5. Displaced Abomasum

Displaced abomasum is a multifactorial disease where the abomasum is dilated as a result of gas accumulation and dislocated to the left (left-displaced abomasum) or to the right (right- displaced abomasum) in the stomach in relation to the normal placing (Ingvartsen, 2006). The passage of feed to the intestines is partly or totally blocked. Approximately 80-90% of the incidences are left-displaced abomasums. The disease is most frequent in high producing cows in early lactation and 80–90% of the cases are seen in the first 4 weeks postpartum (Shaver, 1997). It occurs in approximately 3,5% of dairy cows each year (Goff, 2006a).

Nutrition has been implicated as a major risk factor in the etiology of displaced abomasums, but the precise cause is still unclear. Low feed intake and increased negative energy balance prepartum have been found to increase the risk of displaced abomasums in cows (Goff, 2006a; Ingvartsen,

2006). Cows fed high concentrate diets in early lactation or diets with inadequate particle size are also at increased risk of displaced abomasums (Goff, 2006a). Sudden changes in the diet and rapid increase in concentrate allowance in early lactation are also the risk factors. Some feeds increase the risk of the disease compared to others. Thus, a higher risk of displaced abomasum has been found for cows fed silage compared to those fed hay, probably because silage is often finely chopped. Risk can be almost eliminated if cow eats a kilogram of straw daily (Ingvartsen, 2006).

Displaced abomasum develops if three major conditions are met. The first one is reduced contractility and atony of abomasum which is gas-dilated. The next condition is that the mesentery shall stretch for the abomasum to be able to dislocate, and a third condition is the space in the abdominal cavity (Ingvartsen, 2006). The conditions leading to atony and reduced motility are still unclear, but hypocalcaemia around calving is a possible factor. Reduction in blood Ca concentration around calving results in a reduction of abomasal contractility, which can lead to its atony and dilatation (Shaver, 1997). Hypocalcemic cows after calving have 3,4 to 4,8 times increased risk of development of abomasal displacement (Massey et al., 1993).

Volatile fatty acids (VFA) in the abomasum also have been reported to reduce abomasal motility. Effects of VFA on motility may be exacerbated by low ruminal absorption of VFA during the transition period (Shaver, 1997). Hypocalcaemia also may play a role in this process. Due to reduced feed consumption, slow contractions and inadequate filling of the rumen which therefore does not reach the ventral abdominal wall, empty space appears for movement of abomasum. Then usually more fatty acids escape absorption in the rumen and reach the abomasum. Those VFA, along with hypocalcaemia and accumulated gases, contribute to reducing abomasum contractility and development of atony (Goff & Horst, 1997c). Thus, inadequate feed consumption and insufficient rumen fill with reduced motility and strength of abomasal contractions together contribute to the onset of this disorder (Goff & Horst, 1997c).

Low intake of concentrates during the prepartum period also may increase the risk of left displaced abomasum because absorptive capacity of the ruminal papillae is not increased sufficiently and microbial population of the rumen is not adapted prior to intake of high energy postpartum diets (Shaver, 1997). Too rapid increase of concentrates after calving may reduce roughage intake and potentially increase the risk of displaced abomasum (Ingvartsen, 2006). Uncomplicated ketosis, fatty liver, retained placenta, and metritis are also risk factors for left displaced abomasums (Bobe et al., 2004; Shaver, 1997).

Treatment of displaced abomasum often requires surgical treatment, and deaths are not uncommon so that potential losses due to disease are very high. In addition, cows that recover from abomasal displacement produce about 350 kg less milk during the next month (Shaver, 1997) or 0,8-2,5 kg/day in the rest of lactation (Fourichon et al., 1999). According to Guard (1996) the average losses amounted to about 380 kg of milk annually.

2.6. Milk Fat Depression

Nutrition influences both the quantity and composition of milk fat. In modern dairy production cows are fed diets with high level of concentrates to maximize milk production but such diets often causes drop in milk fat. Condition is known as milk fat depression (MFD) or low-fat milk syndrome. Although milk volume and yield of other milk constituents may not be affected or may be even increased, depression of fat in milk can be a serious economic problem for dairy producers. This syndrome is not a disease, but rather metabolic consequence of attempts to reach the higher milk production in animals.

Milk fat depression refers to a marked reduction in milk fat yield with no change in milk yield or yield of other milk components (Harvatine et al., 2009). It represents a level of milk fat production below the genetic potential of the cow, usually as below 3,2% fat in milk for Holstein, or below 4,2% fat in the milk for a Jersey herd. When problem with MFD exists, the ratio of milk fat/milk protein will be less than 1,0 for Holstein herds. In severe cases the ratio will be less than 0,8. Milk fat yield can be reduced up to 50% and the effect is specific for milk fat (Bauman et al., 2008). The problem has been observed in many feeding situations and dietary conditions, including high level of concentrates and low level of dietary fiber, and diets supplemented with unsaturated oils. The fat content of milk can also be affected by the physical characteristics of the roughage, including grinding and pelleting (Perfield & Bauman, 2005). In diet-induced MFD, the yield of all individual fatty acids is reduced, but the decline is greatest for short- and medium-chain fatty acids that are synthesized de novo in mammary gland (Bauman et al., 2008).

It is generally accepted that induction of MFD requires both an altered rumen environment and the presence of unsaturated fat in the rumen which biohydrogenation pathways are altered (Perfield & Bauman, 2005). Changes in ruminal microbial processes are an essential component for the development of MFD and are often associated with a decrease in rumen pH and a shift in the acetate/propionate ratio.

Historically, there have been several theories proposed to explain the basis for diet-induced MFD. The biohydrogenation theory was proposed by Bauman & Griinari (2001) focusing on the latest research in MFD problem. The authors sugested that "under certain dietary conditions the pathways of rumen biohydrogenation are altered to produce unique fatty acid intermediates which are potent inhibitors of milk fat synthesis". The theory was born after identification of trans-10,cis-12 conjugated linoleic acid (CLA) as a highly potent inhibitor of milk fat synthesis (Bauman & Griinari, 2001). This isomer is a rumen precursor of trans-10 18:1 in biohydrogenation pathway that may arise under certain dietary conditions, including high concentrate/low fiber diets. The biohydrogenation of linoleic acid in rumen involves the formation of trans-11 18:1 and cis-9,trans-11 CLA as the main intermediates. However, under certain dietary situations the other pathways of biohydrogenation can happen and some of the intermediates produced during the process may be potent inhibitors of milk fat synthesis (Bauman & Griinari, 2001). The most extensively studied is trans-10,cis-12 CLA. Minimal quantities of this isomer markedly inhibited milk fat synthesis and a curvilinear reduction in milk fat yield occurred with increasing trans-10,cis-12 CLA in the rumen content (Baumgard et al., 2001).

Isomer trans-10,cis-12 CLA alone does not completely explain the extent of the decrease in milk fat. Additional biohydrogenation intermediates produced in the rumen probably inhibit milk fat synthesis and two of them (trans-9,cis-11 and cis-10,trans-12 CLA) have already been identified (Bauman et al., 2008).

Consequences of MFD are almost exclusively related to losses in yield of milk fat and lower price of milk. As problem is normally reversible, health consequences are usually not a matter of concern. However, MFD inducing diet is usually risky for rumen homeostasis due to high level of carbohydrates and low level of effective fiber which may reduce ruminal pH and cause acidosis (Bergen, 2009). If acidosis is not compensated and drop in ruminal pH is not prevented, the condition may become an important animal health issue.

2.7. Parturient Hypocalcaemia

The lack of Ca in the diet does not lead to any changes in health or production for a long time because it is compensated by mobilization of Ca reserves in the skeleton. The change of Ca content in the blood may not occur until the cows show symptoms of osteoporosis and bone fractures. A unique example, however, is an acute hypocalcaemia in dairy cows that occurs in the periparturient period as a result of sudden loss of great amounts of Ca in colostrum along with temporary dysfunction of the mechanisms of Ca mobilization from bones. From a relative inactivity in the dry period, these mechanisms often fail to achieve full activity quickly enough to maintain normocalcaemia around calving (Horst et al., 1994). More pronounced hypocalcaemia causes progressive neuromuscular dysfunction which is manifested as a specific clinical syndrome known as paresis puerperalis (milk fever, parturient paresis, calving paralysis). More often, however, and in a larger number of cows, hypocalcaemia exists in a subclinical form with very few or no symptoms (subclinical hypocalcaemia).

Hypocalcaemia was defined as the content of total Ca in blood below 2 mmol/L with or without clinical signs of paresis (Oetzel et al., 1988; Radostits et al., 2000), which is roughly equivalent to 1 mmol/L of ionized Ca (Massey et al., 1993; Oetzel, 1996).

Milk fever (MF) is non-febrile disease of adult dairy cows accompanied by general weakness, circulatory collapse and depression of sensation (Oetzel, 1988). It is one of the most common metabolic disorders in intensive dairy production (Horst, 1986) and a typical nutritional disorder (Ender et al., 1971). The disease attacks 5-10% of adult dairy cows in the U.S. and Europe annually, primarily those with high productivity. The classic syndrome of MF occurs immediately before or after parturition with about 87% of cases in the first 48 hours after calving and only about 9% cases before or during the parturition (Oetzel, 1988). The disease is characterized by pronounced hypocalcaemia which is accompanied by hypophosphataemia in most cases, and by more or less serious hypermagnesaemia (Phillippo et al., 1994).

Phenomena like paresis may rarely occur in beef cows, ewes (before or after lambing) and more frequently in dairy goats (Oetzel, 1988). In mares, the disease can occur several weeks after birth and is known as milk tetany or eclampsia (Radostits et al., 2000).

Milk fever has been recognized as a nutritional disorder with various endocrine abnormalities which occur more or less as a secondary phenomenon (Ender et al., 1971). The characteristic of the disease is an acute inability to mobilize Ca from bone (Horst, 1986) despite apparently normal endocrine dynamics in most affected cows (Radostits et al., 2000). The problem seems to lie in the reduced sensitivity of target tissues (bone, kidneys and intestines) on calcitropic hormones due to reduced number of receptors or because of their dysfunction (Horst et al., 1994).

Widely accepted theory for a long time was that MF was caused by a high content of Ca in the diet consumed before calving, as well as by unfavorable ratio of Ca/P. However, further studies emphasized importance of the high content of monovalent cations in the diet, mainly potassium (Block, 1984; Ender et al., 1971; Goff & Horst, 1997b; Horst & Goff, 1997). As a relative excess of cations in the diet affects blood acid-base balance toward alkalosis, the disruption of integrity of tissue receptors is related to metabolic alkalosis that normally exists in cows at the time of calving (Goff, 2000). Thus, the primary cause of MF is considered to be the temporary inability of cow tissues to adequately respond to stimuli generated by calcitropic hormones, mainly parathyroid hormone (PTH) (Goff et al., 1991; Goff, 2000).

In addition to monovalent ions, risk factors also include dietary magnesium (Sansom et al., 1983), dietary P (Curtis et al., 1984), and possibly vitamin D (Goff, 2000; Horst, 1986). Animal factors as the risks for hypocalcaemia include age (Curtis et al., 1984; Dishington, 1974) and breed (Goff, 2000;Oetzel, 1988). Adult cows are at increased risk for the disease. Occurrences are more frequent from the third lactation onwards, reaching the plateau of 12-15% in cows 6-10 years of age (Curtis et al., 1984;Dishington, 1974). The most predisposed breeds for milk fever are Jersey and Guernsey, followed by Holstein and Brown Swiss. In the case of Jerseys possible reasons are considered to be the higher milk production per unit of body weight (Oetzel, 1988) and a higher content of Ca in the colostrum (Goff, 2000). In addition, the number of receptors in the intestine is about 15% lower in Jerseys than in Holsteins (Goff, 2000; Horst et al., 1997).

Depending on the age, breed, feeding regimen and housing conditions, as many as 50-80% of cows can suffer MF annually in some herds (Oetzel, 1988). Concerning that well-timed treatment of MF easily solves the problem and is relatively inexpensive, the disease was not considered as a factor of economic importance in dairy production for a long time (Radostits et al., 2000). However, later findings confirmed its close relationship with other health disorders in the puerperium (Curtis et al., 1985;Massey et al., 1993). Many of these disorders are now considered as complications of hypocalcaemia. They generally occur as a complex, rarely as isolated diseases (Curtis et al., 1985) and include most of metabolic and reproductive disorders in periparturient period: dystocia, retained placenta, metritis, uterine prolapse, ketosis, mastitis and displaced abomasum (Horst et al., 1997). Risk of occurrence of each of them is much higher in hypocalcemic cows (Curtis et al., 1985). A well known physiological effect of hypocalcaemia is atony of smooth and skeletal muscles because their contractility is a function of Ca concentration in extracellular fluid (Ramberg et al., 1984). Most of the possible complications are related specifically to reduced contractility and to atony of smooth muscles of digestive and reproductive tract (Beede, 1995).

Spontaneous recovery of MF is usually not possible and approximately 75% of affected cows die if treatment fails (Oetzel, 1988). Even under normal conditions about 8% of treated cows still die due to various complications, 12% of them are culled (Guard, 1996) and about 25% require another treatment. Due to increased susceptibility to other health disorders and possible complications, the production life of cows that experienced MF was reduced by an average of 3-4 years (Horst et al., 1997).

Subclinical hypocalcaemia influences in the same manner as the clinic one, but to a lesser extent. As is more common in the herd, adverse effects of subclinical hypocalcaemia on herd economy can be equal or even greater than the effects of MF due to its broader influence on feed intake, secondary disease conditions, and milk production during early lactation (Horst et al., 1994).

2.8. Hypomagnesaemia

Magnesium is an essential mineral with many functions in the body but its homeostasis is not hormonally regulated. The concentration of Mg in the blood is dependent exclusively on its absorption from the diet. If Mg secretion in milk and its endogenous losses exceeds absorption from the forestomachs hypomagnesaemia occurs because of the lack of hormonal control. Hypomagnesaemia is common problem in ruminants and may be one of the major health problems of cattle, sheep and goats in large scale production systems especially in temperate climates (Mayland, 1988; Stelletta et al., 2008).

Clinical hypomagnesaemia is also called hypomagnesemic tetany, grass tetany, spring tetany, or lactational tetany. Clinical signs usually occur when the animal is both hypomagnesaemic and hypocalcaemic (Robinson et al., 1989). In its subclinical form, however, hypomagnesaemia may be risk factor for other diseases. Around calving it may be the important risk factor for MF (Sansom et al., 1983; Schonewille et al., 2008).

Hypomagnesaemia may develop due to Mg deficiency in the diet or due to its low utilization in the gut. It may also occur because of a need for increased amounts of Mg during parturition and early lactation (Mayland, 1988). Clinical hypomagnesaemia in cows with plasma Mg concentrations bellow 0,4 mmol/L is manifested as grass tetany (Schonewille et al., 2008). Grass tetany has been investigated extensively, but its complex etiology is not well understood. Large number of factors influences the development of the disease (Robinson et al., 1989). It appears within 2-4 weeks after cattle or sheep have been turned out to rapidly growing pasture. Ewes with twins are more susceptible to grass tetany than are ewes with single lamb (Mayland, 1988). Specifically, older lactating animals consuming lush, intensively fertilized, cool-season grasses during the early spring are the most frequently affected (Robinson et al., 1989). Older animals also have reduced ability to mobilize body reserves of Mg. Lower average environmental temperatures (<14°C) facilitate the onset of the disorder (Mayland, 1988). However, periods of rapid plant growth during any season, resulting in forage of low Mg and high moisture, high nitrogen (N) and K, present dietary conditions that increase the potential for the development of tetany (Robinson et al., 1989). Grass tetany also may develop if animals graze forage that is high in digestible protein, but low in digestible energy (Mayland, 1988).

Grass tetany generally occurs when the dietary intake of total Mg is not particularly low, but factors which increase the animal's requirement for Mg or reduce the availability of dietary Mg are present (Mayland, 1988). The Mg concentrations in forage and subsequently in the blood of cattle are influenced strongly by high amounts of fertilizer K and, to some extent, fertilizer N. Absorption of Mg by plants is reduced by high levels of K in the soil (Mayland, 1988; Robinson et al., 1989). Mg absorption by ruminants is also reduced by a high intake of K (Schonewille et al., 2008), and by high content of Ca and P in the diet (Sansom et al., 1983). High concentrations of N in the forage may additionally decrease the availability of Mg (Mayland, 1988).

Absorption of Mg is dependent on the concentration of Mg in the rumen fluid and the functionality of the Mg active transport process across the rumen wall. The active transport mechanism is critical for the animal when dietary Mg concentration is less than 0,25%. Several factors, such as dietary K, can inhibit Mg absorption by this pathway. At high concentration of Mg in the rumen it will be absorbed by passive transport, and this mechanism is not affected by high K but only by the solubility of Mg in the rumen fluid. This mechanism requires dietary Mg to be not less than 0,35% (Goff, 2006b).

Mg deficiency results in reduced appetite which decreases total nutrient intake (Robinson et al., 1989), thus chronic hypomagnesaemia results in reduced feed intake and milk production. It may progress quickly into acute hypomagnesaemia which terminates in convulsions, coma and death. One-third of animals with clinical symptoms die (Mayland, 1988). Even mild form of hypomagnesaemia (Mg <0,85 mmol/L) significantly reduces the mobilization of Ca from the skeleton around calving (Goff, 2000;Sansom et al., 1983). That is why hypomagnesaemia may be a risk factor for parturient hypocalcaemia and milk fever. According to Sansom et al. (1983) possible reasons for inhibition of Ca mobilization are considered to be:

a. influence on PTH secretion; there are indications that serious Mg deficiency may impair or completely interrupt secretion of PTH,

b. interference with the action of PTH on target tissues; the integrity of interaction between PTH and its receptors is disrupted and activity of initiated enzymes is disabled,

c. interference with the metabolism of vitamin D; the first phase of vitamin D hydroxylation in the liver requires Mg^{2+} ions, while the second phase in the kidney takes place in the presence of PTH.

2.9. Udder Edema

Udder edema occurs sporadically in cows and heifers near parturition, peaking in severity during the immediate prepartum period. Unless complicated, recovery of edema is spontaneous within a few days after calving. However, if edema is sufficiently severe it can interfere with suckling by the calf, milking, and may cause other complications including udder inflammation. Edema is caused by excessive accumulation of fluid underneath the skin and at least some mammary edema is associated with pregnancy and parturition, especially in primigravid heifers (Malven et al., 1983).

Investigations of the causes of udder edema and the development of strategies to reduce its prevalence have progressed in very limited degree during last several decades (Goff, 2006a). Some authors (Schmidt & Schultz, 1959) proposed that severity of udder edema is based on an inherent physiological phenomenon, because a cow may tends to have the same amount of edema each year regardless of the feeding. Feeding program had little if any effect on udder edema (Randalu et al., 1974). "Steaming up" method of feeding dry pregnant cows and high level of concentrates in their diet was initially blamed for high incidences of udder edema. However, investigation found that the amount of edema at calving was not related to grain feeding during the dry period or after calving (Schmidt & Schultz, 1959). Edema also is not correlated with the body condition of the cows (Schmidt & Schultz, 1959). However, addition of Na or K to the diet before calving can increase the incidence and severity of udder edema in dairy cows and first calf heifers (Nestor et al., 1988). In practical terms, high level of NaCl in the feed during the dry period may be one of the major factors in the development of udder edema around calving. Conflicting results from some earlier studies on effect of grain on udder edema could be explained with salt content of the diets. However, KCl as a replacement for NaCl results in edema of about the same severity (Randalu et al., 1974).

Cows with severe edema that requires veterinary treatment are more likely to have additional health problems. Authors found higher culling rates for cows with more severe edema. Thus, there would be a substantial economic benefit to minimizing edema in dairy herds (Dentine & McDaniel, 1984).

2.10. Retained Placenta

Inadequate antioxidant status or "oxidative stress" of the cow contributes to a poorly functioning immune system and increases the risk of mastitis as well as retained placenta (retained fetal membranes, placental retention). Fetal membranes are retained if not expulsed longer than 12 hours after parturition. Se and vitamin E are important dietary antioxidants and their low levels in the diet are associated with a high incidence of mastitis and retained fetal membranes (Goff, 2006a). Subsequently, addition of vitamin E or Se may improve antioxidant status and decrease the incidence of those diseases. The similar effects can be obtained by addition of beta-carotene in the diet. Blood lymphocyte proliferation was higher in cows supplemented with beta-carotene, and phagocytic activity of blood neutrophils was enhanced as well as intracellular killing by blood neutrophils. Therefore, dietary beta-carotene can elevate blood beta-carotene and enhance peripartum host defense mechanisms by enhancing lymphocyte and phagocyte function (Michal et al., 1994).

Hypocalcaemia, among others, may be an important risk factor in the development of retained placenta. Muscle weakening or absence of uterine contractions in hypocalcemic animals does not contribute to the expulsion of fetal membranes (Goff & Horst, 1997c; Oetzel, 1988). In cows with hypocalcaemia placental retention is 3,2 to 4 times more frequent than in normocalcemic cows (Curtis et al., 1985). Hypocalcaemia also delay the physiological involution of uterus and increases the incidence of metritis (Beede, 1995). Moreover, hypocalcaemia is considered as one of the main causes of uterine prolapse, also due to loss of uterine muscles tone.

Cows with retained placenta were 3 times more likely to develop mastitis than animals without retained placenta (Overton & Waldron, 2004). It has been reported that placental retention, metritis and mastitis predisposes dairy cows to foot problems. Moreover, uncomplicated ketosis, retained placenta, metritis, and hypocalcaemia at parturition are also risk factors for left displaced abomasums (Shaver, 1997). Cows with retained placenta or metritis produce 0,3-2,3 kg/day less milk during subsequent lactation (Fourichon et al., 1999). In the case of retention alone, Guard (1996) stated that losses are on average 350 kg of milk annually.

2.11. Metritis

Metritis and endometritis are inflammatory uterine diseases. They frequently occur soon after calving and may severely compromise reproductive performances. Metritis and endometritis refer to the inflammation of the uterus and of its endometrial lining - both conditions are referred to subsequently as metritis (Urton et al., 2005). Younger cows were more likely to have dystocia or assisted deliveries, while older cows were most likely to have retained placenta and metritis (Lewis, 1997).

In some herds, up to 40% of the postpartum cows may be diagnosed with, and treated for uterine infections. However, the exact causes of uterine infections are unknown but are associated with several factors. Cows with dystocia, retained placenta, or stillbirths, and other metabolic disorders are more likely to develop metritis than healthy cows. Impaired immune functions before and after calving seem to predispose cows to severe uterine infections. It has been suggested that the function of neutrophils is impaired in cows that develop uterine infections. Thus, methods for regulating immune function in periparturient cows may have potential for preventing uterine infections. However, prevention of uterine infections is still difficult because the primary causes cannot be defined clearly (Lewis, 1997). Malnutrition influences the ability of the immune system to function, which affects the incidence of diseases such as mastitis and metritis (Goff, 2006a).

Cows suffering metritis exhibit reduced milk yield and reproductive performances (Urton et al., 2005). Few cows die from uterine infections, but affected animals are more likely to be culled for poor reproductive performances. The estimated cost to producers for each cow with metritis was 106 U.S. dollars (Lewis, 1997). Many of the financial costs of metritis are indirect, such as increased days open or predisposition to other diseases, and are thus difficult to measure. As mentioned previously, metritis is one of the diseases that predispose dairy cows to foot problems and left displaced abomasums (Shaver, 1997).

In contrast, the effects of metritis on milk production can be measured immediately, and losses during first four months after calving can be almost 270 kg. Also, besides reduced milk production in sick animals, some pharmaceuticals for treatment of uterine infections contaminate milk with residues, and the milk must be discarded. This is how uterine infections may have an indirect effect on milk production (Lewis, 1997).

3. USE OF BLOOD AND MILK ANALYSES TO EVALUATE NUTRITIONAL AND DISEASE STATUS

If recognition of subclinical diseases is difficult, the condition may be confirmed by analyzing blood, milk or sometimes urine, although some of them are still difficult to diagnose in practice, including rumen acidosis and fatty liver (Ingvartsen, 2006). Blood and milk analyses as tools to evaluate nutritional and disease status of individual animal or whole herd are aimed to help making decisions for improvement of nutritional strategies and production management. However, many problems are associated with interpretation of results of laboratory analyses, including Compton metabolic profile test, as well as other tools for assessment of metabolic status of dairy animals.

3.1. Milk Analyses

Producers can use changes in milk production or in milk composition to monitor the health of their animals, but these tools are not always completely reliable because milk production corresponds poorly with mild or subclinical infectious disease. Nevertheless, measuring energy balance from changes in milk composition, most likely changes in milk fat and protein contents, could provide a cheap and reliable estimator of energy balance. A study clearly showed that there was a strong relationship between energy balance and milk composition under stable feeding conditions (Friggens et al., 2007). Milk composition varies with energy status and was proposed for measuring energy balance on-farm. Ratio milk fat/protein below 1,4 indicates the optimal or positive energy balance, and above it balance is negative (Pehrson, 1996; Zadoks, 2006). During peak lactation many cows with negative energy balance have this ratio in milk even above 2,1 (Pehrson, 1996).

The biological basis for the relationship between fat/protein ratio and energy status of the animal can be found in two physiological features: a) mechanism that maintains milk energy output by increasing milk fat content when yield is compromised due to a deficit in energy supply, and b) decreased milk protein content under negative energy balance (Friggens et al., 2007). Negative energy balance also may be reflected in elevated ketone bodies which are excreted in milk and urine. Milk has about half the ketone level of blood, and is recommended to check milk ketone levels for detection of ketosis.

Milk fat depression is an example of disorder that can be diagnosed exclusively by milk analysis. Specific blood test for MFD diagnosis does not exists, but milk analyses are usually enough to evaluate risk for or presence of MFD in a herd. Every drop in milk fat percentages with no changes in the content of other milk constituents and milk yield can be considered as MFD. In serious cases, diet-induced MFD can result in reduction of milk fat yield of up to 50% or even more (Bauman & Griinari, 2001; Bauman et al., 2008) and represents one of the greatest risks for production economy.

3.2. Blood Analyses

Analyses of blood constituents have long been a matter of concern as a possible tool for assessment of diet adequacy or metabolic and health status of animals. Compton metabolic profile test (CMPT) originally involved the analysis of a set of blood variables from three groups of seven cows, one near peak lactation, and others in midlactation and in late dry period (Kronfeld et al., 1982). The samples are collected at least three times yearly:

summer, autumn and winter, or when nutritional imbalance is expected, using the same procedures and timings (Radostits et al., 2000). Means of variables are calculated for each lactational group in each herd, and this set of means constitutes the metabolic profile. Each group mean is compared to reference ranges determined from corresponding means plus or minus two standard deviations that are based on group-means for all herds (Kronfeld et al., 1982). The test was based on the concept that laboratory measurement of certain components of the blood will reflect the nutritional status of an animal with or without presence of clinical abnormalities. However, the results of research indicate that CMPT may be useful only as an aid in the diagnosis of the nutritional imbalances and production diseases. It must be carefully planed and is still expensive. Laboratories with automated analytical equipment should be available and this is often a major limiting factor (Radostits et al., 2000).

Compton metabolic profile test includes following analyses: blood glucose, packed cell volume, hemoglobin, blood urea nitrogen, serum inorganic phosphate, serum Ca, Mg, K and Na, total serum protein, albumin and globulins, serum Cu and iron (Fe), and plasma NEFA. Obtained results are interpreted with other relevant information taken on the day of sampling related to individual animals or to the herd: age, milk yield, days in milk, concentrates and forage intake, etc (Radostits et al., 2000).

Workers tested the CMPT as a guide to the nutritional status of dairy cows, and found that blood variables were not reliable predictors of energy and nutrients consumption relative to the requirements. Then they supposed that, for prediction of nutritional status, selection of blood variables should be probably different. Those workers have suggested, however, that protein intake may be reflected in blood concentrations of urea, albumin, and hemoglobin. Similarly, blood glucose concentration has attracted attention as an index of energy intake, although results from study to study have been inconsistent (Kronfeld et al., 1982).

Despite limitations and unconformity of metabolic profile tests, blood analyses still can help in diagnostics of some nutritional imbalances. Negative energy balance in postpartum cows can be detected by change in concentration of some blood metabolites (Rushen et al., 2008). As condition that affect immune functions and predispose animals to many metabolic disorders, it is important that negative energy balance is reflected in a higher levels of NEFA and beta-hydroxybutyric acid (BHBA) in the blood. To screen a herd for negative energy balance, Zadoks (2006) recommends having at least 12 animals tested. Testing of NEFA is done 2-14 days before calving and BHBA testing at 2-21 days after calving. If more than 10-15% of animals have NEFA levels above 0,40 mEq/L or BHBA levels above 1,36 mmol/L, the herd is considered to be suffering from excessive negative energy balance. Ketosis and fatty liver are diseases typically associated with negative energy balance for long periods of time. Increased ketone bodies and NEFA in the blood, and decreased blood glucose level are common findings in cows with ketosis and fatty liver. Values of glucose below 2,2 mmol/L are considered subnormal. However, fatty liver in cows currently can be diagnosed only by liver biopsy (Bobe et al., 2004).

The concentration of BHBA in the blood may be taken as a guide for correction of ration supplement for ewes during the final weeks of pregnancy to prevent pregnancy toxemia. It has been recommended to check 10% of the flock and feed is increased if blood concentration of BHBA exceeds 0,8 mmol/L (Morgante, 2004). Besides hypoglycemia and hyperketonaemia, plasma cortisol levels also may be elevated in sheep with pregnancy toxemia (Andrews et al., 1996). Early in the disease, both does and ewes will show a positive test for ketone bodies in the urine (Morgante, 2004).

There are no specific blood variables for detecting metabolic diseases like laminitis, displaced abomasums, udder edema, retained placenta and metritis. However, if these diseases are herd problems, blood metabolic profile should be checked for more than one variable, and carefully used as an aid for development of preventive strategies. When laminitis is a problem, a metabolic profile on both dry and lactating cows as well as springing heifers should include both red and white blood cells count, packed cell volume, Se, Zn, Cu, Fe, blood urea nitrogen, vitamins A and E, and beta-carotene. When retained placenta is a herd problem, blood profile should include serum minerals, Se, vitamin E, and beta-carotene. In individual cases, blood urea nitrogen and packed cell volume should be included. Metritis requires testing metabolic profile on dry and fresh cows which should include white blood cells count, Se, Zn, Cu, Fe, Mg, blood urea nitrogen, vitamins A and E, and beta-carotene.

Use of blood analyses to evaluate degree of hypocalcaemia is sufficient diagnostic test for this disorder. Concentration of Ca may be determined in whole blood, but more often determination is performed in plasma or serum. When there are no disturbances in acid-base status and protein metabolism, ionized and total Ca are strongly correlated and therefore total Ca concentrations in plasma may serve as an acceptable diagnostic value (Oetzel, 1988).

Hypocalcaemia means that the content of total Ca in the blood is below 2 mmol/L, or below 1 mmol/L of ionized Ca (Massey et al., 1993; Oetzel, 1996). Signs of paresis can occur at total Ca level of 1,9 mmol/L, but most of the cows remain on their feet down to 1,0-1,25 mmol/L. A reduction in the plasma Ca concentration at parturition is usually accompanied by hypophosphataemia and hypermagnesaemia (Phillippo et al., 1994). However, this is not always the case, especially if hypocalcaemia is not severe and is not clinically manifested (Crnkic et al., 2010; Joyce et al., 1997;Oetzel et al., 1988).

Blood analysis is also sufficient diagnostic test for hypomagnesaemia. Based on blood serum or plasma Mg levels, animals may be normally magnesemic (0,74-1,23 mmol/L), chronically hypomagnesemic (0,20-0,74 mmol/L) or acutely hypomagnesemic (<0,20 mmol/L) (Mayland, 1988). Sampling and analysis of the blood of several cows within 12 hours after calving is a good indicator of Mg status of the periparturient period. If serum Mg concentrations are less than 0,82 mmol/L in 90% of cows it suggests insufficient dietary Mg absorption (Goff, 2006a).

4. NUTRITIONAL STRATEGIES TO REDUCE METABOLIC DISEASES INCIDENCES

Although metabolic disorders are not easily categorized as to their cause, nutritional strategies have been developed to help prevent many of these disorders (Goff, 2006a). Scientific justification of those measures can be found in many research articles published in scientific journals during the last decades. However, problems in application of the strategies in the field still exist, as well as their limitations and possible negative consequences on milk production economy in certain circumstances. Strategies are based on major factors directly or indirectly increasing the risk of diseases such as overconditioning at calving, excessive mobilization of body fat, low nutrient intake, nutrient or diet specific factors and management and environmental stress (Ingvartsen, 2006). Strategies are expected to be able to reduce morbidity and at the same time improve reproduction and production.

To prevent metabolic disorders in the periparturient period nutritional strategies must start prior to calving. In a survey conducted by Curtis et al. (1985) consumption of nutrients before calving was directly related to the occurrence of metabolic disorders, and directly or indirectly to the occurrence of reproductive disorders after parturition. In most cases, these disorders occur as a complex and many of them are interrelated in their occurrence (Beede, 1995; Curtis et al., 1985). Consequently, strategies to reduce one disease can help preventing others. For example, strategies to reduce liver triglyceride accumulation at calving may decrease incidence of ketosis, etc.

Goff & Horst (1997c) recommended that three basic physiological functions must be maintained during the periparturient period if disease is to be avoided: adaptation of the rumen to lactation diets that are high in energy density, maintenance of a strong immune system, and maintenance of normocalcaemia. Whenever one or more of these functions are impaired, the incidence of both metabolic and infectious diseases is increased. However, measures to maintain these physiological functions must be conducted carefully avoiding other risk factors such as overconditionig and excessive mobilization of body fat around calving.

Metabolic disorders that may appear in the rest of lactation also require adequate prevention strategy to be controlled, including rumen acidosis, laminitis, hypomagnesaemia and milk fat depression.

4.1. Adaptation of The Rumen

Fully adapting the rumen flora to a high starch diet that will be fed after parturition requires about 3 to 4 weeks, and full development of rumen papillae requires about 5 weeks of concentrate feeding (Goff & Horst, 1997c). It is, therefore, important to start increasing concentrates in the diet 3-4 weeks before calving and continue during the first 1-2 weeks after parturition to fully adapt rumen to lactation diet. If fresh cow is abruptly switched to a high starch lactation diet, the risk of developing rumen acidosis exists because the lactate production and accumulation. During the rest of lactation, the most reliable means of preventing rumen acidosis is to apply feeding methods that ensure a more even distribution of feed intake over the day. Feeding TMR only once or twice a day may result in that cows eat an excessive amount of feed during a short period of time, because cows are strongly attracted by the arrival of fresh food (Rushen et al., 2008). Buffering agents such as Na-bicarbonate or alkalinizing agents such as Mg-oxide are added to high concentrate ration to reduce the risk of acidosis (Goff, 2006a). Na-bicarbonate should be supplemented particularly to corn silage-based diets at the rate of approximately 0,8 to 1% of DM (Stone, 2004). Too rapid increase in concentrate allowance during early lactation may reduce roughage intake and increase the risk of not only rumen acidosis, but also displaced abomasums (Ingvartsen, 2006).

A highly significant relationship between forage neutral detergent fiber (NDF) content in the diet and ruminal pH has been found. NRC (2001) recommended 19% of forage NDF as absolute minimum when formulating rations in the field. A system of "physically effective" NDF (peNDF) relates the ability of a feedstuff to stimulate chewing relative to a hypothetical long grass hay containing 100% NDF. The peNDF of a feed is the product of its physically effective fiber (pef) and NDF content. The diet should contain about 22% peNDF to maintain ruminal pH of 6,0 (Stone, 2004).

Gradual adjustment of the rumen to a high concentrate diet can help avoiding serious drop in feed intake around calving and minimize body fat mobilization in early lactation. The plasma NEFA concentration is negatively correlated with DMI and depression in feed intake around the time of calving was largely responsible for fatty liver development (Goff, 2006a). Therefore, all nutritional measures that prevent drop in DMI before parturition may be useful in prevention of fatty liver. However, it has been assumed that increased energy and nutrient density of the diet may assure maintenance of the same intake of nutrients and energy despite lower DMI around calving, and decrease rate of lipid mobilization. Increase in nutrient density during the last 2–3 weeks prepartum by increasing concentrates in the ration has been referred to as "steaming-up" or "close-up" diet (Ingvartsen, 2006). Use these diets should not last for too long because of risk of overfeeding energy and development of obesity. NRC (2001) recommends increasing the energy content of the precalving diet from 5,2 MJ/kg of DM during the "far-off" dry period, to 6,8 MJ/kg of DM for the 3 weeks before calving. This strategy is thought to prepare the cow for the metabolic demands of early lactation and thereby minimize the need for body tissue mobilization (Urton et al., 2005).

As fatty liver and ketosis in cows often occur as a complex, all measures to reduce fatty liver incidence may decrease incidence of ketosis (Grummer, 1993). Moreover, positive effects on ketosis incidence and lipid transport also have been seen when niacin or rumen-protected choline is fed to dry and fresh cows (Goff, 2006a). The treatment of pregnancy toxemia in ewes is usually unsuccessful, therefore the prevention is of key importance to reduce occurrence of the disease. Grain is a high source of available energy. Feeding 0,5-1 kg of grain daily along with high quality hay during the last four to six weeks of pregnancy will help prevent pregnancy toxemia.

Manipulation of the nutritional program of dairy cows affects rumen health, which influences hoof health (Stone, 2004). It is very evident that feeding diets that cause drop in rumen pH will result in increase in laminitis cases. To minimize the drop in rumen pH it is necessary to limit amount of concentrate fed per meal to no more than 3,6 kg and provide fresh feed at the bunk throughout the day. Diet should contain at least 25% total NDF or 19% acid detergent fiber (ADF), with non-fiber carbohydrate (NFC) levels between 35 and 40% or non-structural carbohydrate (NSC) levels between 30 to 35%, and minimize abrupt changes of ration. Adding buffers to diets may help maintain claw integrity as buffers, such as Na-bicarbonate, minimize the drop in rumen pH. Supplements that contain combinations of complexed trace minerals Zn, Mn, Cu and Co positivelly influence claw health, and it has been advised to feed them when lameness is problem in a herd.

Increased energy content of the diet fed during the prepartum period was also associated with decreased incidence of displaced abomasum (Curtis et al., 1985). Risk of displaced abomasum for cows fed silage can be almost eliminated if every cow eats a kilogram of straw daily (Ingvartsen, 2006).

4.2. Maintenance of Immunity

Maintaining of a strong immune system may help to reduce incidence of infectious diseases after parturition including those related to nutritional factors: retained placenta, metritis, mastitis, laminitis, etc. Inadequate nutrition may contribute depression of the immune system that occurs around calving time (Rushen et al., 2008) when stress of parturition and metabolic challenges experienced by the dairy cow at the onset of milk production impair immune cell function (Goff, 2006a).

Dietary factor that may influence immune functions include vitamin E and Se as the most important vitamin and mineral related to immunity. To a lesser extent vitamin A and beta-carotene, Cu and Zn may play a role (Zadoks, 2006). Metabolic conditions related to parturition also may exacerbate immunosuppression, including negative energy balance (Zadoks, 2006), hypocalcaemia (Ganjkhanlou et al., 2010), and hyperketonemia (Overton & Waldron, 2004). Therefore, strategies to maintain strong immune system must include feeding adequate or increased levels of all dietary components that influence immunity, as well as prevention of metabolic disorders that may exacerbate immunosuppression.

Most measures explained in the preceding section related to adaptation of the rumen to lactation diets may help maintain immune functions throughout reduction in negative energy balance and hyperketonemia. Measures to reduce hypocalcaemia around calving are presented in the following section. Feeding vitamin E and Se have long been associated with reducing incidences of retained placentas and uterine infections. Increased protein content of the diet fed during the prepartum period was also associated with decreased incidences of retained placenta (Curtis et al., 1985).

Although feeding is generally not related to the incidence and severity of udder edema, some dietary measures can help prevention of the problem. Restriction of the salt NaCl in the diet reduces the severity of udder edema (Randalu et al., 1974). Restriction or exclusion of other Na or K sources from the diet is also recommended until the levels of dietary Na and K are not in line with dietary recommendations. NRC (2001) recommends at least 0,1% Na in dietary DM for close-up dry cow diet and at least 1% K. As membrane permeability may be a primary cause of udder edema, vitamin E supplementation precalving may be a supportive preventive measure (Thomas et al., 1990). More recently, feeding anionic diets or diets with additional antioxidants have shown some promise in reducing udder edema (Goff, 2006a). Prevention of udder edema is important as condition has temporary effects of pain and stress for the cow, and may increase risk of udder inflammation and additional health problems.

4.3. Prevention of Parturient Hypocalcaemia

Concerning effects of hypocalcaemia on many physiological functions it is obvious that health, production and reproduction can be compromised in hypocalcemic cows even in the absence of clinical signs of paresis (Oetzel, 1996). Therefore, measures to prevent decline of blood Ca in cows after calving can improve milk production in herds that apparently have no problem with this disease. Prevention aims to increase the mobilization of Ca from the skeleton, or its absorption in the gastrointestinal tract, or both. The most important nutritional measures for prevention of hypocalcaemia include manipulation of minerals in the diet which is nutritionally balanced in accordance with requirements of the cows several weeks before calving. This method of prevention is based on limiting the consumption of Ca or increasing the content of anions in the diet.

4.3.1. Anionic Diet

An anionic, or acidic, diet is one supplemented with anionic salts to provide more anions (Cl- and S2-) relative to the cations (Na+ and K+). It has a negative dietary cation-anion difference (DCAD), calculated as mEq/kg of DM = (Na + K) – (Cl + S). It is well known that anionic diets help maintain blood Ca at parturition and prevent milk fever when fed to cows during the last several weeks of pregnancy (Block, 1984). Feeding an anionic diet before parturition has been advised if the incidence of milk fever in a herd exceeds 10%, and also when it is desired to improve the health status and production in herds in which MF is not a serious problem (Horst et al., 1994; Oetzel, 1993).

Diet with excess of anions, mainly Cl- and S2-, relative to the cations Na+ and K+ acidifies the body and is therefore considered as acidogenic (Goff & Horst, 1997b). The two PTH-dependent functions, including bone resorption and production of 1,25-(OH)2D, were enhanced after feeding acidogenic diet which contributes to lower incidence of hypocalcaemia after calving (Goff, 2000). However, the diets of dry cows in the field are always more or less alkalogenic and have a positive DCAD. When feeding diets with positive DCAD the acid-base balance tends to metabolic alkalosis, and vice versa. Formulation of acidogenic diet is only possible by supplementation with chloride and sulfate in a quantity that provides a relative excess of Cl- and S2- in the diet. Manipulation of dietary cations and anions, however, is limited by metabolic requirements and by tolerance levels of minerals in the diet.

The literature offered a dozen different formulas to calculate DCAD, which include some or all dietary minerals with or without the use of coefficients related to their utilization from the diet or the degree of influence on acid-base status:

A. (Na+K) – (Cl+S) Ender et al. (1971)

B. (Na+K) – (Cl+S+P) Lomba et al. (1978)

C. (Na+K) – Cl Gaynor et al. (1989)

D. (Na+K+Ca+0,3P) – (Cl+S)Goff et al. (1991)

E. (Na+K+0,38Ca+0,3Mg) – (Cl+0,6S+0,5P)Goff (1992)

F. (Na+K+0,38Ca+0,25P+0,3Mg) – (Cl-0,6S) Goff (1992)

G. (Na+K+0,38Ca+0,3Mg) – (Cl-0,6S)Horst et al. (1997)

H. (Na+K+0,15Ca+0,15Mg) – (Cl+0,2S+0,3P)Oetzel (2000)

I. (Na+K+0,15Ca+0,15Mg) – (Cl+0,25S+0,5P)Goff (2000)

J. (Na+K) – (Cl+0,6S)Goff et al. (2004)

Equations A, B, and C imply the complete absorption of each dietary element that can be considered accurate only in the case of equation C. Equations E and G take into account the coefficients of utilization of those elements that are not completely utilized. Equations H and I could be considered biologically and functionally most accurate because they include the degree of influence of individual ions on acid-base status, and are physiologically the most relevant (Horst & Goff, 1997; Goff, 2000).

Equation F can be used to assess the risk of developing MF. All the elements that increase the risk of MF are on the left side, while those who decrease the risk are on the right side of equation. Equation D can serve the same purpose (Goff et al., 1991). The last equation J, (Na + K) – (Cl + 0,6S), was the most recently proposed by Goff et al. (2004).

Oetzel (1991) collected data from 75 published experiments and determined the risk factors of diet in the development of MF using meta-analysis technique. Comparing the three different equations to calculate DCAD he found that the equation (Na + K) - (Cl + S) was strongly correlated with the appearance of MF. The author in the same study well justified the inclusion of S in the calculation of DCAD. Charbonneau et al. (2006) used the same technique but found that the equation (Na + K) – (Cl + 0,6S) was the most highly associated with clinical milk fever (R2 = 0,44) and urinary pH (R2 = 0,85).

Mineral content in feed is usually expressed as a percentage so it is necessary to convert it in mEq/kg for calculation of DCAD. Equivalent weight (g/Eq) of each element is calculated by dividing the relative atomic weight (Ar) with its valence. For example, Ar of S is 32,06. Dividing 32,06 by 2 we get the equivalent weight of S which is 16,03 g/Eq. Factor for conversion of percentage in mEq/kg is calculated by dividing the 10.000 with equivalent weight of the element. For S it is 10.000/16,03 = 623,83 (Table 1). Multiplying this factor with percentage of the element in dietary DM its content is expressed in mEq/kg DM. When this is calculated for each element in the equation, then the sum of cations is subtracted from the sum of anions and DCAD value is obtained. The example of calculating DCAD of alfalfa hay and rapeseed meal using equation (Na + K) - (Cl + S) is given in Table 2.

Table 1. Mineral elements in the diet used for calculation of DCAD

Element	A_r	Valence	Equivalent weight (g/Eq)	Conversion factor (from % in mEq/kg)
Na	22,99	1	22,99	434,98
K	39,09	1	39,09	255,77
Ca	40,08	2	20,04	499,00
Mg	24,31	2	12,16	822,88
Cl	35,45	1	35,45	282,06
S	32,06	2	16,03	623,83
P	30,97	3	10,32	968,56

Table 2. Example of calculating DCAD in alfalfa hay and rapeseed meal

Alfalfa hay	K	Na	Cl	S
Content in DM, %	1,71	0,12	0,38	0,28
Factor for multiplication (F)	255,77	434,98	282,06	623,83
Content in mEq/kg DM (% × F)	437,37	52,20	107,18	174,67
DCAD mEq/kg DM [(Na+K) – (Cl+S)]	489,57 – 281,85 = + 207,72			
Rapeseed meal				
Content in DM, %	1,36	0,10	0,11	1,25
Factor for multiplication (F)	255,77	434,98	282,06	623,83
Content in mEq/kg DM (% × F)	347,85	43,50	31,03	779,79
DCAD mEq/kg DM [(Na+K) – (Cl+S)]	391,35 – 810,82 = - 419,47			

The example in Table 2 clearly shows the difference in the alkali-acid production potential of the two feeds depending on their mineral content. A positive DCAD of alfalfa hay is result of the relatively high K content. Extremely negative DCAD of rapeseed meal is result of high S content.

Most diets commonly used for dry cows have DCAD value of +100 to +250 mEq/kg DM (Oetzel, 2000) and even up to +500 mEq/kg DM (Pehrson et al., 1999). By choosing appropriate feeds DCAD can be reduced, but not sufficiently to influence Ca metabolism. It is therefore necessary to enrich the diet with Cl and S by adding anionic salts in a quantity that provides a negative DCAD. That is, indeed, a standard diet that contains all essential nutrients required by dry cow and was further enriched with sources of Cl and S in amount that causes a mild metabolic acidosis.

Acidogenic, or anionic mineral salts are chlorides and sulfates with a high content of Cl and S, not containing Na and K. Alkalogenic salts are, on the other hand, Na and K salts that contain organic anion which is metabolized and are also called cationic salts (Na2CO3, NaHCO3, K2CO3, KHCO3, etc). They have the effect opposite to anionic salts and are undesirable in the diet of dry cows.

To formulate an anionic diet three chlorides and three sulfates are commonly used (Table 3). Two or more of them are usually added depending on mineral content of the diet. Ca and Mg salts also serve to meet requirements in these two minerals whose content in anionic diet should be higher than in standard diet (NRC, 2001). Ammonium salts, besides Cl and S, are also a source of non-protein nitrogen.

Acid production potential or acidifying activity of certain salt depends on preferential absorption of anions in relation to cations which the salt is consisted of (Goff et al., 1991). Phosphates, for example, have a weak acidifying activity as they are absorbed only slightly better than the corresponding cations (Horst et al., 1997). NaCl is neutral salt as both elements are absorbed completely and none is metabolized, so it does not affect neither DCAD nor acid-base status.

Anionic salts are not harmless substances and require caution for the use and handling them. The main limiting factors are bitter and salty taste and potential toxicity of higher doses. NH4Cl is considered the most toxic, followed by CaCl2. The combination of salts is the best solution because it reduces the possibility of harmful effects (Oetzel, 2000). The most acceptable method of administration is mixing salts in the feed.

An alternative to the anionic salts is adding mineral acids directly into a diet. Very potent in causing systemic acidosis is hydrochloric acid (HCl), while sulfuric acid (H2SO4) is less efficient (Goff &Horst, 1997a). An additional advantage of mineral acids is that cows prefer diets with acidic taste rather than bitter-salty taste of anionic salts. Goff & Horst (1998) found that efficient prevention of MF can be achieved with only 1,5 Eq of HCl in the diet, and that with no adverse consequences up to 2,5 Eq of the acid can be fed daily. Moreover, the addition of HCl in the diet had, unlike anionic salts, a beneficial effect on the consumption of feed. For safety, it is not advisable to use pure acid or keep it on farm, but rather as commercial products in the form of acidified feeds where the acid is mixed with some feed as a carrier. Byproducts of fermentation are usually used for this purpose, and also soybean meal or sugar beet pulp (Goff & Horst, 1997a, 1998).

Table 3. Magnesium sulfate heptahydrate (Epsom salt), 2magnesium chloride hexahydrate, 3calcium sulfate dihydrate (gypsum), 4calcium chloride dihydrate, 5ammonium sulfate, 6amonium chloride, Mr – relative molecular weight of a substance (sum of atomic weights of all atoms in the molecule), Val – valence, g/Eq – equivalent weight (Mr/valence of cation), DCAD = (Na+K) – (Cl+S).Table 3.Characteristics and chemical composition of anionic salts

Salt	M_r	Val	g/Eq	DM, %	% of DM					DCAD
					Ca	Mg	N	Cl	S	mEq/kg DM
$MgSO_4x7H_2O$ [1]	246,51	2	123,26	48,83		20,20			26,63	-16.613
$MgCl_2x6H_2O$ [2]	203,33	2	101,67	46,83		25,53		74,46		-21.002
$CaSO_4x2H_2O$ [3]	172,18	2	86,09	79,07	29,44				23,55	-14.691
$CaCl_2x2H_2O$ [4]	147,02	2	73,51	75,49	36,11			63,88		-18.018
$(NH_4)_2SO_4$ [5]	132,14	2	66,07	100,0			21,20		24,26	-15.134
NH_4Cl [6]	53,49	1	53,49	100,0			26,19	66,27		-18.692

[1]magnesium sulfate heptahydrate (Epsom salt), [2]magnesium chloride hexahydrate, [3]calcium sulfate dihydrate (gypsum), [4]calcium chloride dihydrate, [5]ammonium sulfate, [6]amonium chloride, M_r – relative molecular weight of a substance (sum of atomic weights of all atoms in the molecule), Val – valence, g/Eq – equivalent weight (M_r/valence of cation), DCAD = (Na+K) – (Cl+S).

Feeding anionic diet few weeks before calving significantly reduces the incidences of MF and subclinical hypocalcaemia in the herd, and improves milk production and reproductive performances of cows in the subsequent lactation (Beede et al., 1991; Block, 1984; DeGroot et al., 2010). Metabolic acidosis caused by anionic diet is subclinical, mild and compensated, and with no significant impact on animal health, but may influence some physiological functions if lasts long enough. Increased sensitivity of PTH receptors in renal tissue contributes to increased production of 1,25-(OH)2D before parturition (Gaynor et al., 1989; Goff et al., 1991). Cows fed anionic diet have a higher content of 1,25-(OH)2D in plasma, although changes in secretion of PTH does not occur (Phillippo et al., 1994). The active form of vitamin D takes part in osteolysis with PTH (Horst et al., 1994; Horst, 1986) and stimulates Ca reabsorption in renal tubules (Goff, 1992). Since 1,25-(OH)2D is important for the functioning of both Ca compensation mechanisms, change in its secretion is considered as an important effect of anionic diet.

Oetzel (1993) recommended the application of anionic salts in the herds with high incidence of MF (>10%) if not possible to formulate low Ca diet, or to improve production and reproduction in the herd which apparently does not have high incidence of MF. Adding anions in the diet is, therefore, to be considered not only for preventing MF, but also for prevention of subclinical hypocalcaemia which is responsible for the frequent occurrence of other metabolic and reproductive disorders in puerperium (Horst et al., 1994). Beede (1992) states that even in well-kept herds, where cows are in proper condition with no major problems with metabolic disorders, an additional 250-500 kg of milk can be achieved by using anionic salts in diets before parturition probably because lower incidence of subclinical hypocalcaemia.

Based on results of earlier studies, Horst et al. (1997) found that the most effective diets for prevention of MF are those with DCAD from -50 to -100 mEq/kg DM, while others state that it is between -50 and -150 (Beede, 1995), or -200 (Horst et al., 1994). However, Beede et al. (1991) achieved a good preventive effect with DCAD -250 mEq/kg, and Goff et al. (1991) with -230 mEq/kg. Too low DCAD, irrespective of the amount of added salts, can reduce feed consumption because too strong metabolic acidosis which depresses the appetite in animals (Goff & Horst, 1997b). Preventive effect of anionic diet on MF incidence occurs only if DCAD is below -40 mEq/kg DM in most cases, regardless of the type and quantity of added salts.

A risk for many health and reproductive disorders in puerperium is increased in cows with hypocalcaemia (Curtis et al., 1985; Massey et al., 1993). The prevention of hypocalcaemia can, in addition to better health status of cows in puerperium, improve milk production and increase reproduction performances in the herd (Beede et al., 1991; Block, 1984; DeGroot et al., 2010). Comparing the milk production in healthy cows with those who suffered MF, Block (1984) found greater milk production by about 14% in healthy animals. The same author found that cows fed anionic diet in the dry period produced on average 486 kg (7,3%) more milk than those fed cationic diet. Beede et al. (1991) conducted a field experiment with 510 cows and showed three positive effects of anionic diet: enhanced Ca metabolism in periparturient period, increased milk production in subsequent lactation, and better reproductive performances. Cows fed anionic diet before parturition produced 327 kg more milk (about 3,6%) than control cows.

Because of unpleasant taste, anionic salts can have depressing effect on feed intake (Joyce et al., 1997;Vagnoni & Oetzel, 1998), but this is minimal if salts are properly used (Block, 1984; Moore et al., 2000; Oetzel et al., 1988). Up to 300 mEq of anions/kg DM (3-3,5 Eq of salts per day per cow) can be added to the diet without depressing effect on feed intake (Horst et al., 1994; Oetzel, 2000). Intake also could be influenced by the type of salt because more acidic salts have less favorable taste. Feed intake depression was at least expressed with $MgSO4$, and much more with other salts, mainly $CaCl2$ and $NH4Cl$ (Oetzel & Barmore, 1993). Last two also have much higher acidifying activity in relation to $MgSO4$.

The best and easiest method of application of anionic salts is mixing in TMR which effectively masks their taste. In the conventional feeding system application of anionic salts is difficult, but still possible. The salts should be manually mixed in silage or haylage, or in concentrates (Oetzel & Barmore, 1993). In the latter case, salts should be mixed in sufficient amount of concentrates containing palatable feeds, at least 2,3-2,6 kg (Oetzel & Barmore, 1993; Pehrson et al., 1999).

Oetzel (2000) finds that preventive effect of anionic salts can be achieved if cows consume anionic diet at least 5 days before calving, while others state that at least 10 days is necessary. Since the expected and actual date of calving are matter of discrepancies in practice, it is advisable to start feeding salts 3-4 weeks before expected parturition, at least 2-3 weeks, to ensure that most of the cows consume the diet at least 10 days (Beede 1992, 1995).

NRC (2001) gives fairly broad recommendations for the content of Ca in anionic diet for dry cows, from 0,6 to 1,5% DM, which is not difficult to formulate when anionic Ca salts are used. Anionic salts should not be fed if Ca intake is below 50 g/day (Oetzel, 1993). As hypomagnesaemia negatively affects Ca metabolism at the time of calving, many authors emphasize the importance of adequate content of Mg in dry cow diet (Sansom et al., 1983; Thilsing-Hansen et al., 2002). Recommendation has increased from earlier 0,20-0,25% to the current 0,35-0,40% DM either in standard or in anionic diet (NRC, 2001). Increasing dietary Mg up to 0,40% DM has no negative consequences so, even if not prove useful, poses no risk or any practical problem in the formulation of a diet when anionic Mg salts are used (Wang & Beede, 1992). On the other hand, the Mg level of 0,4% DM has been also set as the maximum tolerable (NRC, 2001).

If we assume that DCAD in efficient anionic diet should be around -100 mEq/kg DM, it means that the initial DCAD should not be greater than +200 to +250 (Horst et al., 1997). Therefore, the diet must be formulated using feeds with lower content of K, such as corn silage, grass hay, etc. If necessary, part of the forage with high content of K can be replaced with concentrates rich in fiber, such as brewers' grains, sugar beet pulp, malt sprouts, etc. The most common problem is grass silage and alfalfa silage because they can contain more than 3% K. Most cereal grains have almost neutral DCAD, while it is more negative only in rapeseed meal and brewers grains primarily due to high content of S (NRC, 2001).

In field conditions, a marked decrease in feed intake may occur when anionic salts are included in the diet for the first time, therefore, adjusting cows should last at least three days (Oetzel, 2000)

The most accurate biological indicator of the degree of acidification of the body is pH of urine (Vagnoni & Oetzel, 1998). Urinary pH in cows was decreased linearly with decreasing DCAD (Charbonneau et al.,

2006; Ganjkhanlou et al., 2010). The optimum pH of urine to prevent puerperal hypocalcaemia in Holsteins fed anionic diet is 6 - 6,5 (Horst & Goff, 1997) or wider, 6 - 7 (Moore et al., 2000). In Jersey cows, however, the pH of urine is necessary to reduce to 5,8 - 6,2. If the pH is lower than the specified, amount of anionic salts should be reduced and vice versa.

4.3.2. Low Calcium Diet

Maintenance requirements of the cow and requirements of fetus during high pregnancy are usually satisfied with 35-45 g Ca/day (NRC, 2001). The content of Ca in the diet is often beyond that, and the requirements in this case are met mainly by passive absorption, while mechanisms of mobilization of body reserves and active absorption of Ca in the intestine are suppressed. Inactivity of these mechanisms in the dry period makes cows difficult to adapt to the sudden loss of Ca in colostrum at the moment of mammary gland activation. A few days required for starting these mechanisms causes temporary crisis in Ca homeostasis that results in decline of Ca in the blood (Goff, 1992).

When Ca content in the diet is below the minimum requirement, the animal is dependent on Ca mobilization from bone with simultaneous increase of active Ca absorption in the gut (Horst, 1986). These mechanisms can be activated before parturition and maintained active by constant stimulation up to the critical moment of parturition. Restriction of Ca intake to below 20 g/day caused negative balance of Ca and stimulates the secretion of PTH which increases tubular Ca reabsorption, bone resorption and production of 1,25-(OH)2D (Goff, 1992; Horst et al., 1997). This allows the cow to use Ca more efficiently from the diet and from body reserves immediately after calving (Goff, 1992). Introduction of low Ca diet leads to a slight decrease of Ca and P concentrations in plasma which returns to the base level in 3-4 days due to simultaneous increase in secretion of PTH that remained increased up to calving time (Goings et al., 1974). Prolonged exposure of the tissues to elevated PTH levels after feeding low Ca diet can overcome tissue resistance to PTH that might be induced by high dietary K (Goff, 2006a). This is the way to reduce or avoid period of adaptation to sudden losses of Ca which usually takes several days (Goff, 1992). Rations formulated to contain less than 20 grams of Ca proved to be very effective in preventing MF and significantly reduce the occurrence of hypocalcaemia too (Goings et al., 1971, 1974; Wiggers et al., 1975). For the best effect, restrictions of Ca consumption in the dry period should last at last 7-10 days (Goings et al., 1974; Wiggers et al., 1975) and be folowed by a high Ca diet immediately after calving (Horst et al., 1994; Oetzel, 2000). Cows fed in this way at the moment of calving mobilize about 10 g of Ca from the skeleton daily that is enough to prevent the occurrence of MF (Wiggers et al., 1975). Increasing

Ca in the diet above daily needs and manipulation of its content within these limits does not affect the blood Ca status if other dietary minerals are maintained unchanged (Goff & Horst, 1997b).

Feeding low Ca diet to dry cows is considered the traditional way of preventing MF. For this reason, the use of feeds rich in Ca is usually avoided and the diet is formulated with corn silage and grass hay. Consumption of Ca can be reduced to 50-60 g/day in this way what often gives good results in the field (Goff, 1992; Horst et al., 1997). However, significant stimulation of parathyroid gland and complete preventive effect can be achieved only if Ca consumption is restricted to below 20 g/day. The formulation of such diet is, unfortunately, unrealistic in practice and that is the reason for its limited application (Goff, 1992). This type of diet has no adverse effects on production in the subsequent lactation (Goings et al., 1974), but the problem is, in addition to impracticality, that it may not last longer than 2-3 weeks because negative balance of Ca can be too exhausting for body reserves (Van Saun & Sniffen, 1996).

In the strategy for prevention of MF, Horst et al. (1997) recommended measures depending on the baseline cation-anion difference of the diet (Fig. 1). If it is less than 250 mEq/kg DM, the use of chloride and sulfate is justified without danger of low feed intake. If this value is above 250 mEq/kg DM it is necessary to consider other preventive measures, such as low Ca diet, short-term administration of oral Ca salts or some pharmacologic therapy (vitamin D analogs and active metabolites, PTH injections, etc.).

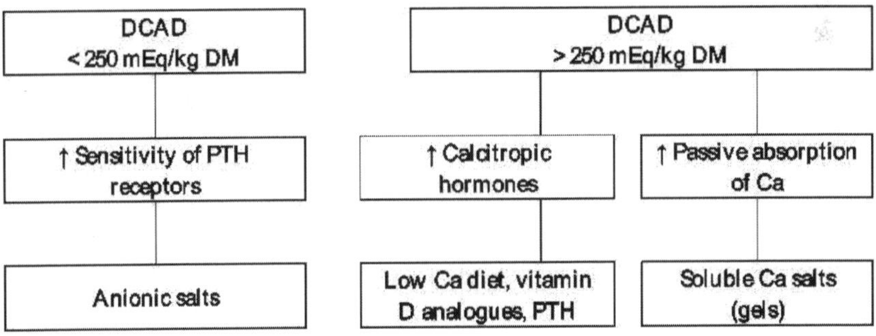

Figure 1. Strategy for preventing milk fever (adapted from Horst et al., 1997)

Thilsing-Hansen et al. (2002) rewieved research conducted over the past 50 years and concluded that any measures for preventing MF, even if used under ideal conditions, rarely reach preventive effect of 100%. The same authors calculated that efficiency of low Ca diets is best, reaching 80-100%, and anionic diets efficiency was 65-80%.

4.4. Prevention of Hypomagnesaemia

Commonly used parameter to characterize the grass tetany potential of forage is the ratio [K/(Ca+Mg)] (Mayland, 1988). Forages containing less than 0,2% Mg and a "tetany ratio" [K/(Ca+Mg)] greater than 2,2 have increased risk of inducing grass tetany (Crawford et al., 1998). The fertilizers containing N and K are the most important factors increasing [K/(Ca+Mg)] ratio in forages. Key step in nutritional prevention of hypomagnesaemia is finding a forage with lower ratio [K/(Ca+Mg)], or modify this ratio in the diet by adding more Ca and Mg supplements. Commercial grade MgO, MgC12, MgCO3 and MgSO4 are good sources. As hypomagnesaemia may induce hypocalcaemia, including Ca in supplements to prevent Mg tetany may have a beneficial effect (Robinson et al., 1989).

NRC (2001) recommends Mg level of 0,35-0,40% DM in the diet for dry cows, although studies showed no benefit from increasing dietary Mg above 0.2% DM when diets contain less than 3% K (Wang & Beede, 1992). Van Saun & Sniffen (1996) recommended increasing dietary Mg above 0,20% when the content of K exceeds 1,2 to 1,5% DM. However, Goff (2006a) recommends that Mg content of the close-up dry cow ration and the early lactation ration should be between 0,35 and 0,40% as insurance against the possibility that the active transport processes for Mg absorption are impaired. In pregnant cows high Ca intake decreased utilization of dietary Mg and increases its excretion in urine (Sansom et al., 1983) that can be even worsened in metabolic acidosis (Wang & Beede, 1992). So it seems logical to increase Mg content in anionic close-up diet because it usually contains more Ca.

4.5. Prevention of Milk Fat Depression

As milk fat depression has been observed over a wide range of feeding situations, this problem on dairy farms remains one of the more challenging tasks within overall nutritional management of dairy cows. Diets high in concentrates and low in fiber, and diets supplemented with plant or fish oil are the most often associated with MFD, but many other dietary factors also can affect milk fat synthesis including those which are able to alter rumen environment and those related to supply of polyunsaturated fatty acids (PUFA). Factors that can alter rumen environment include low level of physically effective NDF, feed particle size, total fiber in the diet, starch and non-structural carbohydrates, feeding pattern etc. On the other side, factors related to supply of PUFA are variation in fat content and fatty acid composition of feed ingredients, amount and availability of PUFA, and also feeding pattern.

Factor that alter rumen environment are first to consider in nutritional strategy to prevent or solve the problem of MFD, and those who affect rumen pH are the most important. Lower then normal ruminal pH, even without signs of acidosis, causes the change in bacterial population favoring those that have alternative pathways of biohydrogenation of dietary fatty acids. Inclusion of adequate level of so-called ''effective fiber'' and appropriate buffers in lactation diets can prevent drop in ruminal pH and markedly decrease MFD (Bergen, 2009). Forages, including long stemmed hay, are the main sources of effective fiber, thus the best method to maintain an adequate fat percentage in milk is to feed a balanced ration with adequate forage. The low levels of effective fiber may result from overfeeding of concentrates or the lack of forage, from consumption of large amounts of lush pasture and from silage or haylage that is too finely chopped (Perfield & Bauman, 2005).

Additional management practices to maintain a stable milk fat percentage in dairy herds include regular feeding of diet without abrupt changes, as well as feeding buffers such as Na-bicarbonate and/or Mg-oxide. Buffers are particularly useful when more than 5,5 kg of concentrate is fed per feeding and when frequent changes in diet are made.

Adding different sources of fat in dairy cow rations is a practice which has been favored when it is necessary to increase the energy consumption. However, higher amounts of fat can inhibit the activity of rumen bacteria and reduce the efficiency of fiber digestion, thus leading to a reduction in milk fat content (Schroeder et al., 2004). Although unsaturated vegetable oils can have a number of positive effects on fatty acid composition of milk fat, a negative effect on fat content must not be neglected due to the fact that the fat in milk is one of the main determinants of the price of milk and profits that farmers make. High levels of unsaturated oil in the diet can reduce fat content in milk along with possibly other adverse effects on production performances, such as drop in milk yield, decreased protein content in milk, low feed intake, etc. (Shingfield et al., 2006). Loor et al. (2002) found that the upper limit for dietary supplementation of unsaturated vegetable oils is 3,5% DM without serious nefative consequences on production. If fish oil is the supplement, it is at most 1% DM (Donovan et al., 2000).

REFERENCES

1. A. H. Andrews, V. E. Holland-Howes, J. I. D. Wilkinson, 1996Naturally Occurring Pregnancy Toxaemia in the Ewe and Treatment with Recombinant Bovine SomatotropinSmall Ruminant Research232-3January 1997), 1911970921-4488

2. L. G. Baird, C. K. W. Muelling, 2009Risk Factors, Pathogenesis and Prevention of Subclinical Laminitis in Dairy Cows. Proceedings of the CanWest Veterinary Conference, 110Alberta, Canada, October 17-20, 2009

3. D. E. Bauman, J. M. Griinari, 2001Regulation and Nutritional Manipulation of Milk Fat: Low-Fat Milk SyndromeLivestock Production Science701-2July 2001), 15291871-1413

4. D. E. Bauman, I. I. J. W. Perfield, K. J. Harvatine, L. H. Baumgard, 2008Regulation of Fat Synthesis by Conjugated Linoleic Acid: Lactation and the Ruminant ModelJournal of Nutrition1382February 2008), 4034090022-3166

5. L. H. Baumgard, J. K. Sangster, D. E. Bauman, 2001Milk Fat Synthesis in Dairy Cows is Progressively Reduced by Increasing Supplemental Amounts of trans-10,cis-12 Conjugated Linoleic Acid (CLA).Journal of Nutrition, 1316June 2001), 176417690022-3166

6. D. K. Beede, C. Wang, G. A. Donovan, L. F. Archibald, W. K. Sanchez, 1991Dietary Cation-Anion Difference (Electrolite Balance) in Late Pregnancy. Proceedings of the 28th Florida Dairy Production Conference, 98104Gainesville, FL, April 9-10, 1991

7. D. K. Beede, 1995Macromineral Element Nutrition for the Transition Cow: Practical Implications and Strategies. Proceedings of The Tri-State Dairy Nutrition Conference, 175188Fort Wayne, IN, USA, May 11-13, 1995

8. D. K. Beede, 1992The DCAD Concept: Transition Rations for Dry Pregnant Cows. Feedstuffs, 6453December 1992), 12190014-9624

9. W. G. Bergen, 2009Milk-Fat Depression and Lipid Repartitioning in Lactating Dairy CowsJournal of Nutrition1395May 2009), 8268270022-3166

10. S. J. Bertics, R. R. Grummer, C. Cadorniga-Valino, E. E. Stoddard, 1992Effect of Prepartum Dry Matter Intake on Liver Triglyceride Concentration and Early Lactation.Journal of Dairy Science757July 1992), 191419220022-0302

11. E. Block, 1984Manipulation Dietary Anions and Cations for Prepartum Dairy Cows to Reduce Incidence of Milk Fever. Journal of Dairy Science, 6712December 1984), 293929480022-0302

12. G. Bobe, J. W. Young, D. C. Beitz, 2004Invited Review: Pathology, Etiology, Prevention, and Treatment of Fatty Liver in Dairy Cows.Journal of Dairy Science8710October 2004), 310531240022-0302

13. E. Bramley, I. J. Lean, W. J. Fulkerson, M. A. Stevenson, A. R. Rabie, N. D. Costa, 2008The Definition of Acidosis in Dairy Herds Predominantly Fed on Pasture and ConcentratesJournal of Dairy Science911January 2008), 3083210022-0302

14. E. Charbonneau, D. Pellerin, G. R. Oetzel, 2006Impact of Lowering Dietary Cation-Anion Difference in Nonlactating Dairy Cows: A Meta-Analysis.Journal of Dairy Science892February 2006), 5375480022-0302

15. R. J. Crawford Jr, M. D. Massie, D. A. Sleper, H. E. Mayland, 1998Use of an Experimental High-Magnesium Tall Fescue to Reduce Grass Tetany in Cattle. Journal of Production Agriculture114October-December 1998), 4914960890-8524

16. C. Crnkic, S. Muratovic, S. Piplica, A. Kavazovic, S. Kutlaca, 2010Blood Plasma Mineral Profile and Health Status in Postpartum Cows Fed an Anionic Diet Before ParturitionTurkish Journal of Veterinary and Animal Science, 343June 2010), 2552601300-0128

17. C. R. Curtis, H. N. Erb, C. J. Sniffen, R. D. Smith, D. S. Kronfeld, 1985Path Analysis of Dry Period Nutrition, Postpartum Metabolic and Reproductive Disorders, and Mastitis in Holstein Cows.Journal of Dairy Science689September 1985), 234723600022-0302

18. C. R. Curtis, H. N. Erb, C. J. Sniffen, R. D. Smith, 1984Epidemiology of Parturient Paresis: Predisposing Factors with Emphasis on Dry Cow Feeding and Management.Journal of Dairy Science674April 1984), 8178250022-0302

19. M. A. De Groot, E. Block, P. D. French, 2010Effect of Prepartum Anionic Supplementation on Periparturient Feed Intake, Health, and Milk ProductionJournal of Dairy Science9311November 2010), 526952790022-0302abstract)

20. M. R. Dentine, B. T. Mc Daniel, 1984Associations of Subjective Udder Edema Scores and Descriptive Trait Codes for Udder Types.Journal of Dairy Science671January 1984), 2082150022-0302

21. I. W. Dishington, 1974The Role of Age on the Induction of Hypocalcaemic Paresis Puerperalis in Dairy Cows. Nordisk Veterinaermedicin, 263March 1974), 2052100029-1579

22. D. C. Donovan, D. J. Schingoethe, R. J. Baer, J. Ryali, A. R. Hippen, S. T. Franklin, 2000Influence of Dietary Fish Oil on Conjugated Linoleic Acid and Other Fatty Acids in Milk Fat from Lactating Dairy Cows.Journal of Dairy Science8311November 2000), 262026280022-0302

23. F. Ender, I. W. Dishington, A. Helgebostad, 1971Calcium Balance Studies in Dairy Cows Under Experimental Induction and Prevention of Hypocalcaemic Paresis PuerperalisZeitschrift fur Tierphysiologie, Tierernahrung und Futtermittelkunde, 281-5January-December 1971), 2332560044-3565

24. C. Fourichon, H. Seegers, N. Bareille, F. Beaudeau, 1999Effects of Disease on Milk Production in the Dairy Cow: A Review.Preventive Veterinary Medicine411June 1999), 1350167-5877

25. N. C. Friggens, C. Ridder, P. Løvendahl, 2007On the Use of Milk Composition Measures to Predict the Energy Balance of Dairy Cows.Journal of Dairy Science9012December 2007), 545354670022-0302

26. M. Ganjkhanlou, A. Nikkhah, A. Zali, 2010Effect of Dietary Cation-Anion Balance on Milk Production and Blood Mineral of Holstein Cows During the Last Two Months of PregnancyAfrican Journal of Biotechnology936September 2010), 598359881684-5315

27. P. J. Gaynor, F. J. Mueller, J. K. Miller, N. Ramsey, J. P. Goff, R. L. Horst, 1989Parturient Hypocalcaemia in Jersey Cows Fed Alfalfa Haylage Based Diets With Different Cation to Anion Ratios. Journal of Dairy Science, 7210October 1989), 252525310022-0302

28. J. P. Goff, R. L. Horst, F. J. Mueller, J. K. Miller, G. A. Kiess, H. H. Dowlen, 1991Addition of Chloride to a Prepartal Diet High in Cations Increases 1,25-dihydroxyvitamin D Response to Hypocalcaemia Preventing Milk FeverJournal of Dairy Science7411November 1991), 386338710022-0302

29. J. P. Goff, R. Ruiz, R. L. Horst, 2004Relative Acidifying Activity of Anionic Salts Commonly Used to Prevent Milk Fever.Journal of Dairy Science875May 2004), 124512550022-0302

30. J. P. Goff, 1992Cation-Anion Difference of Diets and Its Influence on Milk Fever and Subsequent Lactation: the Good and the Bad News.

Proceedings of Cornell Nutrition Conference For Feed Manufacturers, 1481590885-7687Rochester, N.Y, USA, October 13-15, 1992

31. J. P. Goff, 2000Pathophysiology of Calcium and Phosphorus Disorders.Veterinary Clinics of North America: Food Animal Practice, 162July 2000), 3193370749-0720

32. J. P. Goff, 2006aMajor Advances in Our Understanding of Nutritional Influences on Bovine Health.Journal of Dairy Science894April 2006), 129213010022-0302

33. J. P. Goff, 2006bMineral Disorders of the Transition Period: Origin And Control. Proceedings of XXIV World Buiatrics Congress, 192-90362-340-7France, October 15-19, 2006

34. J. P. Goff, R. L. Horst, 1997aComparison of Sulfuric Acid with Hydrochloric Acid as a Source of Acidifying Anions. Journal of Animal Science, 75Suppl.1, (January 1997), 980021-8812

35. J. P. Goff, R. L. Horst, 1997bEffects of the Addition of Potassium or Sodium, but not Calcium, to Prepartum Rations on Milk Fever in Dairy CowsJournal of Dairy Science801January 1997), 1761860022-0302

36. J. P. Goff, R. L. Horst, 1997cPhysiological Changes at Parturition and their Relationship to Metabolic Disorders.Journal of Dairy Science807July 1997), 126012680022-0302

37. J. P. Goff, R. L. Horst, 1998Use of Hydrochloric Acid as a Source of Anions for Prevention of Milk Fever.Journal of Dairy Science8111November 1998), 287428800022-0302

38. R. L. Goings, N. L. Jacobson, E. T. Littledike, 1971Responses of Cows to a Very Low Calcium Intake at Parturition. Journal of Dairy Science, 545May 1971), 7917920022-0302

39. R. L. Goings, N. L. Jacobson, D. C. Beitz, E. T. Littledike, K. D. Wiggers, 1974Prevention of Parturient Paresis by a Prepartum Calcium-Deficient Diet.Journal of Dairy Science5710October 1974), 118411880022-0302

40. L. E. Green, V. J. Hedges, Y. H. Schukken, R. W. Blowey, A. J. Packington, 2002The Impact of Clinical Lameness on the Milk Yield of Dairy Cows.Journal of Dairy Science859September 2002), 225022560022-0302

41. R. R. Grummer, 1993Etiology of Lipid-Related Metabolic Disorders in Periparturient Dairy Cows.Journal of Dairy Science7612December 1993), 388238960022-0302

42. C. L. Guard, 1996Fresh Cow Problems are Costy; Culling Hurts the Most. Hoard's Dairyman, 14180018-2885

43. K. J. Harvatine, Y. R. Boisclair, D. E. Bauman, 2009Recent Advances in the Regulation of Milk Fat SynthesisAnimal31January 2009), 40541751-7311

44. R. L. Horst, 1986Regulation of Calcium and Phosphorus Homeostasis in Dairy Cow. Journal Of Dairy Science692February 1986), 6046160022-0302

45. R. L. Horst, J. P. Goff, 1997Milk Fever and Dietary Potassium. Proceedings of Cornell Nutrition Conference for Feed Manufacturers, 1811890885-7687Rochester, NY, USA, October 21-23, 1997

46. R. L. Horst, J. P. Goff, T. A. Reinhardt, D. R. Buxton, 1997Strategies for Preventing Milk Fever in Dairy Cattle.Journal of Dairy Science807July 1997), 126912800022-0302

47. K. L. Ingvartsen, 2006Feeding- and Management-Related Diseases in the Transition Cow: Physiological Adaptations Around Calving and Strategies to Reduce Feeding-Related Diseases. Animal Feed Science and Technology, 1263-4March 2006), 1752130377-8401

48. E. H. Johnson, K. Al-Habsi, E. Kaplan, A. Srikandakumar, I. T. Kadim, K. Annamalai, R. Al-Busaidy, O. Mahgoub, 2004Caprine Hepatic Lipidosis Induced Through the Intake of Low Levels of Dietary CobaltThe Veterinary Journal1682September 2004), 1741791090-0233

49. P. W. Joyce, W. K. Sanchez, J. P. Goff, 1997Effect of Anionic Salts in Prepartum Diets Based on Alfalfa.Journal of Dairy Science8011November 1997), 286628750022-0302

50. D. S. Kronfeld, S. Donoghue, R. L. Copp, F. M. Stearns, R. H. Engle, 1982Nutritional Status of Dairy Cows Indicated by Analysis of Blood.Journal of Dairy Science6510October 1982), 192519330022-0302

51. G. S. Lewis, 1997Uterine Health and Disorders.Journal of Dairy Science805May 1997), 9849940022-0302

52. D. N. Logue, J. E. Offer, 2001Guest Editorial: The Effect of Forage Type on Foot Health in Dairy Heifers. The Veterinary Journal, 1621July 2001), 781090-0233

53. F. Lomba, E. T. Chauvaux, L. Lengele, V. Bienfet, 1978Calcium Digestibility in Cows as Influenced by the Excess of Alkaline Ions Over Stable Acid Ions in Their Diets.The British Journal of Nutrition393May 1978), 4254290007-1145

54. J. J. Loor, L. E. Quinlan, A. B. P. A. Bandara, J. H. Herbein, 2002Distribution of trans-vaccenic acid and cis-9,trans-11-conjugated linoleic acid (rumenic acid) in blood plasma lipid fractions and secretion in milk fat of Jersey cows fed canola or soybean oil. Animal Research, 512March-April 2002), 1191341627-3583

55. P. V. Malven, R. E. Erb, M. F. D'amico, T. S. Stewart, B. P. Chew, 1983Factors Associated with Edema of the Mammary Gland in Primigravid Dairy Heifers.Journal of Dairy Science662February 1983), 2462520022-0302

56. C. D. Massey, C. Wang, G. A. Donovan, D. K. Beede, 1993Hypocalcaemia at Parturition as a Risk Factor for Left Displacement of the Abomasum in Dairy Cows. Journal of the American Veterinary Medical Association, 2036September 1993), 8528530003-1488

57. H. F. Mayland, 1988Grass Tetany. In: The Ruminant Animal; Digestive Physiology and Nutrition, D.C. Church, (Ed.), 511531Prentice Hall, 0-88133-740-4Cliffs, N.J., USA

58. J. J. Michal, L. R. Heirman, T. S. Wong, B. P. Chew, M. Frlgg, L. Volker, 1994Modulatory Effects of Dietary Beta Carotene on Blood and Mammary Leukocyte Function in Periparturient Dairy Cows.Journal of Dairy Science775May 1994), 140814210022-0302

59. S. J. Moore, M. J. Van de Haar, B. K. Sharma, T. E. Pilbeam, D. K. Beede, H. F. Bucholtz, J. S. Liesman, R. L. Horst, J. P. Goff, 2000Effects of Altering Dietary Cation-Anion Difference on Calcium and Energy Metabolism in Peripartum Cows.Journal of Dairy Science839September 2000), 209521040022-0302

60. M. Morgante, 2004Digestive Disturbances and Metabolic-Nutritional Disorders.In: Dairy sheep nutrition. G. Pulina (Ed.), 165191CAB International, 0-85199-681-7UK.

61. K. E. Nestor Jr, R. W. Hemken, R. J. Harmon, Influence of Sodium Chloride and Potassium Bicarbonate on Udder Edema and Selected Blood Parameters.Journal of Dairy Science712February 19883663720022-0302

62. NRC,2001Nutrient Requirement Of Dairy Cattle.th rev.ed., National Academy Press, 0-30906-997-1DC, USA

63. G. R. Oetzel, 1988Parturient Paresis and Hypocalcaemia in Ruminant Livestock. Veterinary Clinics of North America: Food Animal Practice, 42July 1988), 3515640749-0720

64. G. R. Oetzel, 1991Meta-Analisis of Nutritional Risk Factor for Milk Fever in Dairy Cattle. Journal of Dairy Science, 7411November 1991), 390039120022-0302

65. G. R. Oetzel, 1993Use of Anionic Salts for Prevention Milk Fever in Dairy Cattle. Compendium On Continuing Education For The Practicing Veterinarian, 158August 1993), 113811460193-1903

66. G. R. Oetzel, 1996Effect of Calcium Chloride Gel Treatment in Dairy Cows On Incidence of Periparturient Diseases.Journal of the American Veterinary Medical Association2095September 1996), 9589610003-1488

67. G. R. Oetzel, 2000Management of Dry Cows for the Prevention of Milk Fever and Other Mineral Disorders.Veterinary Clinics of North America: Food Animal Practice, 162July 2000), 3693860749-0720

68. G. R. Oetzel, J. A. Barmore, 1993Intake of a Concentrate Mixture Containing Various Anionic Salts Fed to Pregnant, Nonlactating Dairy Cows.Journal of Dairy Science766Juny 1993), 161716230022-0302

69. G. R. Oetzel, J. D. Olson, C. R. Curtis, M. J. Fettman, 1988Ammonium Chloride and Ammonium Sulfate for Prevention of Parturient Paresis in Dairy Cows.Journal of Dairy Science7112December 1988), 330233090022-0302

70. T. R. Overton, M. R. Waldron, 2004Nutritional Management of Transition Dairy Cows: Strategies to Optimize Metabolic HealthJournal of Dairy Science87E. Suppl.), (July 2004), E105E1190022-0302

71. B. Pehrson, 1996Milk Analysis as an Indicator of the Nutritional and Disease Status of Dairy Cow. In: Recent Advances in Animal Nutrition, P.C. Garnsworthy, J. Wiseman, W. Haresign, (Eds.), 113129Nottingham University Press, 1-89767-603-4

72. B. Pehrson, C. Svensson, I. Gruvaeus, M. Virkki, 1999The Influence of Acidic Diets on the Acid-Base Balance of Dry Cows and the Effect of Fertilization on the Mineral Content of Grass.Journal of Dairy Science826June 1999), 131013160022-0302

73. I. I. J. W. Perfield, D. E. Bauman, 2005Current Theories and Recent Advances in the Biology of Milk Fat Depression. Proceedings of Cornell Nutrition Conference for Feed Manufacturers, 951069-99825-829-4

74. M. Phillippo, G. W. Reid, I. M. Nevison, 1994Parturient Hypocalcaemia in Dairy Cows: Effects of Dietary Acidity on Plasma Minerals and Calciotrophic Hormones.Research in Veterinary Science563May 1994), 3033090034-5288

75. O. M. Radostits, C. C. Gay, D. C. Blood, K. W. Hinchcliff, 2000Veterinary Medicine: A Textbook of the Diseases of Cattle, Sheep, Pigs, Goats and Horses.th Edition. 14171476WB Saunders, 0-70202-604-2UK

76. C. J. Ramberg Jr, E. K. Johnson, R. D. Fargo, D. S. Kronfeld, 1984Calcium Homeostasis in Cows with Special Reference to Parturient Hypocalcaemia. American Journal of Physiology, 2465May 1984), R698R7040363-6119

77. W. E. Randalu, R. W. Hemken, L. S. Bull, L. W. Douglas, 1974Effect of Dietary Sodium and Potassium on Udder Edema in Holstein Heifers. Journal Of Dairy Science, 574April 1974), 4724750022-0302

78. D. L. Robinson, L. C. Kappel, J. A. Boling, 1989Management Practices to Overcome the Incidence of Grass Tetany.Journal of Animal Science6712December 1989), 347034840021-8812

79. J. Rushen, A. M. de Passillé, Keyserlingk. M. A. G. von, D. M. Weary, 2008The Welfare of CattleAnimal Welfare Series, 5C.Phillips (Ed.), Springer, 978-1-40206-558-3Dordrecht, The Netherlands

80. B. F. Sansom, R. Manston, M. J. Vagg, 1983Magnesium and Milk FeverVeterinary Record11219May 1983), 4474492042-7670

81. C. Schlumbohm, J. Harmeyer, 2003Hypocalcaemia Reduces Endogenous Glucose Production in Hyperketonemic Sheep. Journal of Dairy Science, 866June 2003), 195319620022-0302

82. G. H. Schmidt, L. H. Schultz, 1959Effect of Three Levels of Grain Feeding During the Dry Period on the Incidence of Ketosis, Severity of Udder Edema, and Subsequent Milk Production of Dairy Cows.Journal Of Dairy Science421January 1959), 1701790022-0302

83. J. T. Schonewille, H. Everts, S. Jittakhot, A. C. Beynen, 2008Quantitative Prediction of Magnesium Absorption in Dairy CowsJournal of Dairy Science911January 2008), 2712780022-0302

84. G. F. Schroeder, G. A. Gagliostro, F. Bargo, J. E. Delahoy, L. D. Muller, 2004Effects of Fat Supplementation on Milk Production and Composition by Dairy Cows on Pasture: A ReviewLivestock Production Science861-3March 2004), 1181871-1413

85. R. D. Shaver, 1997Nutritional Risk Factors in the Etiology of Left Displaced Abomasum in Dairy Cows: A Review.Journal of Dairy Science8010October 1997), 244924530022-0302

86. C. Stelletta, M. Gianesella, M. Morgante, 2008Metabolic and Nutritional Diseases.In: Dairy Goats Feeding And Nutrition. A.Cannas, G.Pulina (Eds.), 263288CAB International, 1-84593-348-6UK

87. W. C. Stone, 2004Nutritional Approaches to Minimize Subacute Ruminal Acidosis and Laminitis in Dairy CattleJournal of Dairy Science87E. Suppl., (July 2004), E13E260022-0302

88. T. Thilsing-Hansen, R. J. Jorgensen, S. Ostergaard, 2002Milk Fever Control Principles: A Review.Acta Veterinaria Scandinavica431March 2002), 1191751-0147

89. D. G. Thomas, J. K. Miller, F. J. Mueller, F. C. Madsen, 1990Udder Edema Reduced by Prepartum Vitamin E Supplementation. Journal of Dairy Science, 73Suppl.1, 2710022-0302

90. G. Urton, Keyserlingk. M. A. G. von, D. M. Weary, 2005Feeding Behavior Identifies Dairy Cows at Risk for Metritis.Journal of Dairy Science888August 2005), 284328490022-0302

91. D. B. Vagnoni, G. R. Oetzel, 1998Effect of Dietary Cation-Anion Difference on the Acid-Base Status of Dry Cows. Journal of Dairy Science816June 1998), 164316520022-0302

92. R. J. Van Saun, C. J. Sniffen, 1996Nutritional Management of the Pregnant Dairy Cow to Optimize Health, Lactation and Reproductive PerformanceAnimal Feed Science and Technology591-3June 1996), 13260377-8401

93. C. Wang, D. K. Beede, 1992Effects of Diet Magnesium on Acid-Base Status and Calcium Metabolism of Dry Cows Fed Acidogenic Salts.Journal of Dairy Science753March 1992), 8298360022-0302

94. K. D. Wiggers, D. K. Nelson, N. L. Jacobson, 1975Prevetion of Parturient Paresis by a Low-Calcium Diet Prepartum: A Field Study. Journal of Dairy Science, 583March 1975), 4304310022-0302

95. R. N. Zadoks, 2006The Role of Nutrition in Udder Health. Proceedings of High Plains Dairy Conference, 115128Albuquerque, NM, March 17, 2006

CHAPTER 2

Metabolic Profile and Inflammatory Responses in Dairy Cows with Left Displaced Abomasum Kept under Small-Scaled Farm Conditions

Fenja klevenhusen 1,†, elke humer 1,†, barbara metzler-zebeli 2,†, leopold podstatzky-lichtenstein 3, thomas wittek 4 and qendrim zebeli 1,*

1Department of Farm Animals and Veterinary Public Health, Institute of Animal Nutrition and Functional Plant Compounds, Vetmeduni Vienna, Veterinaerplatz 1, Vienna 1210, Austria

2Department of Farm Animals and Veterinary Public Health, University Clinic for Swine, Vetmeduni Vienna, Veterinaerplatz 1, Vienna 1210, Austria

3Veterinary Clinic Dr. Leopold Podstatzky-Lichtenstein, Wels 4600, Austria

4Department of Farm Animals and Veterinary Public Health, University Clinic for Ruminants, Vetmeduni Vienna, Veterinaerplatz 1, 1210 Vienna, Austria

SIMPLE SUMMARY

This research established an association between lactation number and milk production and metabolic and inflammatory responses in high-producing dairy cows affected by left abomasal displacement in small-scaled dairy farms. The study showed metabolic alterations, liver damage, and inflammation in the sick cows, which were further exacerbated with increasing lactation number and milk yield of the cows.

ABSTRACT

Left displaced abomasum (LDA) is a severe metabolic disease of cattle with a strong negative impact on production efficiency of dairy farms. Metabolic and inflammatory alterations associated with this disease have been reported in earlier studies, conducted mostly in large dairy farms. This research aimed to: (1) evaluate metabolic and inflammatory responses in dairy cows affected by LDA in small-scaled dairy farms; and (2) establish an association between lactation number and milk production with the outcome of metabolic variables. The cows with LDA had lower serum calcium (Ca), but greater concentrations of non-esterified fatty acids (NEFA) and beta-hydroxy-butyrate (BHBA), in particular when lactation number was >2. Cows with LDA showed elevated levels of aspartate aminotransferase, glutamate dehydrogenase, and serum amyloid A (SAA), regardless of lactation number. In addition, this study revealed strong associations between milk yield and the alteration of metabolic profile but not with inflammation in the sick cows. Results indicate metabolic alterations, liver damage, and inflammation in LDA cows kept under small-scale farm conditions. Furthermore, the data suggest exacerbation of metabolic profile and Ca metabolism but not of inflammation and liver health with increasing lactation number and milk yield in cows affected by LDA.

KEYWORDS

metabolic disease; acute phase response; periparturient dairy cow; milk yield; cow welfare

1. INTRODUCTION

The transition period (*i.e.*, the period occurring two weeks prepartum through two to four weeks postpartum) is marked by extensive dietary, metabolic, endocrine, and immunological changes in dairy cows [1,2]. This period is also characterized by a high incidence of metabolic diseases, including left displaced abomasum (LDA) in post-partum dairy cows [3]. Displaced abomasum is a severe disease that affects between 0.8% and 6.3% of dairy cows, mostly within 4 weeks postpartum [4]. This disease is characterized by varying degrees of abomasal distension and displacement [5] and is preferably treated by surgical correction. Displaced abomasum has strong negative impacts on overall production efficiency of the afflicted farms mainly due to high related costs (approximately $494 per case) [6], and direct association to decreased herd health status and fertility [5,7].

Although the mechanisms behind LDA are not completely known, research has reported significant associations between the incidence of LDA and a negative energy balance (NEB) prepartum, often reflected in elevated circulating non-esterified fatty acids (NEFA) and beta-hydroxy-butyrate (BHBA) concentrations [1,8]. Other predisposing factors for LDA seem to include the first four weeks of lactation [9], genetics [10,11], neurological disorders linked to hypomotility/atony of the abomasum [9,12], metabolic disorders (e.g., ketosis, off-feed), as well as other concomitant diseases, such as endometritis, mastitis, and claw disorders [9], which lead to activation of the acute phase response (APR), characterized by elevated levels of blood acute phase proteins (APP) in the affected cows [12]. Indeed, multiple studies have reported increased blood concentrations of the APP haptoglobin (Hp) and serum amyloid A (SAA) in cows with LDA [11,12,13,14,15].

Despite this large volume of research around LDA, there is a paucity of information, in particular, with regards to potential metabolic and inflammatory conditions associating LDA in small-scale dairy farms. As shown by Stengärde *et al.*[16] the incidence of LDA increases with the size of the herd. Small-scale farms have fewer cows and more individual management options than medium or large size farms [16,17], possibly paying more attention to individual cows. On the other hand, the small-sized farms may have less access to modern technology, likely impairing their capability to properly manage LDA.

The role of lactation number and of milk production in the incidence of LDA and, most importantly, in metabolic and APR variables has not yet been clarified. For example, Cameron *et al.* [8] observed a decreased incidence rate of LDA when lactation number increased, but other field studies have reported an opposite trend [18], or no significant effect of the lactation number on the frequency of LDA [19]. Contradictory observations have also been reported on the relationship of milk yield and LDA [9], and no data are available to which extent these factors are able to modulate metabolic and APR responses. Therefore, this research aimed to evaluate metabolic and APR responses in high-producing dairy cows affected by LDA in small-scale dairy farms additionally taking into account the lactation number and milk production of LDA cows as compared to their healthy counterparts.

2. MATERIAL AND METHODS

2.1. Animals and Treatments

Fourteen small-scaled dairy farms in Austria having between 14 to 88 cows/farm (median value: 55 cows/farm) were included in this longitudinal study. All of them were enrolled in the milk recording program of the Association of Austrian Cattle Breeders. All experimental procedures were in compliance with the institutional guidelines and national legislation for animal use in research and approved by the institutional animal care committee. All owners gave informed consent for cows to be included in this study and undergo the testing procedures. Complete information regarding farm conditions, health records, and milk production and composition of the cows was recovered. Average milk production of farms involved in the study ranged from 7765 to 12,073 kg/cow/year with 3.8% to 4.5% fat, and 3.2% to 3.6% protein in the milk. From a total of 20 dairy cows from these farms with LDA blood samples were taken. All farms had a free stalls system and feeding consisted of partial mixed ration and supplementary concentrate mixture for dry and fresh cows during close-up and early postpartum periods, respectively.

Samples of other 20 postparturient clinically-healthy dairy cows of possibly close lactation stage, number, and breed with the LDA cows were used to contrast the results of serum analyses of cows with LDA. In the LDA group, 16 cows were Holstein Friesian and four cows were Simmental breed. The LDA cows had on average 2.9 lactations (four cows in first lactation, six cows in second lactation, three cows in third lactation, four cows in the fourth lactation, and one cow in the fifth, sixth, and seventh lactation each). Cows of the control group (17 Holstein cows and three Simmental cows) had on average of 2.5 lactations (two cows in first lactation, nine cows in second lactation, seven cows in third lactation, and two cows in the fourth lactation). To evaluate the effect of cow's lactation number on metabolic variables in cows affected with LDA or not, cows were classified in two different lactation groups: cows with ≤2 lactations (n = 10 in LDA group and n = 11 in the control group) or cows with >2 lactations (n = 10 in LDA group and n = 9 in the control group).

Cows were diagnosed with LDA from day 1 to 30 (on average: 14 d) postpartum and were immediately submitted to surgery. The clinical diagnose of LDA was confirmed by this surgery intervention via right-flank laparotomy under local analgesia (Procamidor 20 mg/mL, Richter Pharma AG, Wels, Austria). 25 mL were injected twice subcutaneously and intramuscularly in the incision line. The abomasal body was displaced to the left dorsal abdominal quadrant between the left body wall and the rumen. After abdominal exploration and confirmation of LDA, the abomasum was decompressed, repositioned and fixated by omentopexy to its normal anatomical position. All cows underwent surgical correction of LDA while standing. To ensure a rapid and smooth recovery from the surgery, the cows affected with LDA were administered 50 mL of a combination of penicillin and streptomycin intramuscularly (Pen Strep, Univet, Irland) and intravenous infusion of 500 mL glucose solution (Glucose B. Braun 200 mg/mL Infusionslösung, B. Braun Melsungen AG, Melsungen, Germany). All cows that underwent LDA surgery recovered fully.

2.2. Blood Sampling

Blood samples were collected from the coccygeal vein shortly prior to surgery for the LDA cows. Blood samples from healthy cows were collected on average 16 ± 5.2 d (mean ± SD) postpartum, a sampling time being comparable with 14 ± 8.9 d postpartum for LDA cows. Serum 10 mL vacutainer tubes (Vacuette, Greiner bio-one, Kremsmünster, Austria) were used to collect blood samples. After collection, blood samples were shortly allowed to clot and serum was separated by centrifuging at 3000× *g* at 4 °C for 20 min. Serum samples were stored at −20 °C until analysis.

2.3. Serum Analyses

Concentrations of SAA and Hp in serum were determined by commercially-available ELISA kits according to the methods described previously [15,20]. In brief, serum samples for SAA were initially diluted 1:500, and samples with optical density values above the range of the standard curve were diluted further (1:400 or 1:250) and re-analyzed. No dilution of serum was necessary for measurement of Hp. All samples were tested in duplicate and the optical density values were read on the iMark microplate absorbance reader (Bio-Rad Laboratories GmbH, Vienna, Austria) at 450 nm. The intra-assay variation of all APP assays was controlled by coefficient of variations limits ≤10%.

Serum concentrations of BHBA, NEFA, Calcium (Ca), Phosphorus (P), blood urea nitrogen (BUN), cholesterol, and liver enzymes such as aspartate aminotransferase (AST), glutamate dehydrogenase (GLDH) and gamma-glutamyltransferase (GGT) were measured at the platform for laboratory diagnostics of the University of Veterinary Medicine, Vienna, using standard enzymatic colorimetric analysis with a fully automated autoanalyzer for clinical chemistry (Cobas 6000/c501; Roche Diagnostics GmbH, Vienna, Austria), as described recently [21].

2.4. Milk Data

Milk yields were recorded and milk samples were collected from the morning and afternoon milking of cows starting from 45 d postpartum (*i.e.*, on average four weeks after the surgery) during at least three occasions in a week interval. Milk samples of healthy cows were collected at similar time postpartum with DA cows. Milk samples were analyzed for fat, protein, and somatic cell counts (SCC) by Milkoscan (Foss Electric, Hillerød, Denmark).

2.5. Statistical Analyses

Data were first analyzed for normality using Shapiro-Wilk test. Except for NEFA and NEFA/Chol ratio, all other variables were normally distributed by this test ($p < 0.05$). Therefore, NEFA and NEFA/Chol were log-transformed for the parametric analysis. The analysis of variance was done using the MIXED procedure of SAS. An overall model was fitted to take into consideration the fixed effect of lactation number group (*i.e.*, ≤2 lactations and >2 lactations), health status, and their two-way interaction as well as the random effect of farm. The general statistical model was as follows:

$$Yijk = \mu + Pi + Hj + Fk + (PH)ij + eijk \quad (1)$$

in which Yijk represents the observation (*i.e.*, continuous variable such as serum variables or milk data) on cow k at lactation number group i of health status j and farm k; μ is the overall mean, Pi represents the fixed effect of the i-th lactation number group, i = 1, 2; Hj represents the fixed effect of the j-th health status, j = 1, 2; (PH)ij is the interaction between lactation number group and health status, and Fk represents the random effect of the k-th farm, k = 1, 2, … 14, and e the experimental error [22].

Degrees of freedom were approximated by the method of Kenward-Roger (ddfm = kr). Least square mean (LSM) and the respective standard error of the mean (SEM) were computed. Significance was declared at $p \leq 0.05$, while a tendency was considered at $0.05 < p \leq 0.10$. To evaluate the effect of cow's milk yield on metabolic variables, linear regression analyses using PROC REG of SAS were performed, as follows:

$$Y = \alpha 0 + \beta X + e \tag{2}$$

in which Y represents the response variable; α_0 is the intercept, β is the regression slope and X describes the milk yield of cows, defined as explanatory variable, and e states for residuals. Only significant and tendentially significant ($p < 0.10$) relationships were considered. p-value and R^2 were computed and used to evaluate the goodness of fit. All statistical analyses were carried out with SAS software (SAS Institute Inc., Cary, NC, USA, version 9.2).

3. RESULTS

3.1. Serum Variables

Data showed major alterations in several serum metabolic variables in cows affected by LDA (Table 1). These cows had dramatic increases of NEFA, NEFA/cholesterol ratio, and BHBA concentrations compared to healthy cows ($p < 0.001$). Aside from the health status, lactation number of the cows affected several serum variables. Accordingly, cows with ≤2 lactations had lower serum NEFA and BHBA concentrations of 25% and 50% respectively than cows with >2 lactations did ($p < 0.05$). Serum BUN was lower in cows with ≤2 lactation compared to cows >2 lactations ($p < 0.05$), and there was an interaction between the two factors on this variable ($p = 0.077$). Serum cholesterol was affected neither by LDA nor by lactation number.

Serum minerals were not affected by lactation number, but LDA cows had reduced serum Ca concentrations compared to healthy cows ($p < 0.001$). There was an interaction between lactation number and serum Ca ($p < 0.05$), indicating a greater difference in Ca levels between sick and healthy cows with increasing lactation number. Likewise, the Ca/P ratio was lower in LDA cows ($p < 0.05$).

Table 1. Serum variables of cows with left abomasal displacement (LDA) or healthy cows (HC) differing in lactation number.

Serum Variables [1]	≤2 Lactations		>2 Lactations		SEM [2]	*p*-value [3]		
	LDA	HC	LDA	HC		Lact	Status	Int
Metabolites								
Cholesterol (mg/dL)	88.2	91.2	84.5	98.4	8.40	0.873	0.451	0.625
NEFA (mmol/L)	1.17	0.52	1.56	0.69	0.095	0.043	<0.001	0.434
BHBA (mmol/L)	2.11	0.74	4.67	1.00	0.428	0.028	<0.001	0.072
NEFA/cholesterol	0.018	0.006	0.023	0.007	0.002	0.258	<0.001	0.402
BUN (mg/dL)	23.7	35.2	35.6	35.9	2.23	0.046	0.981	0.077
Minerals								
Calcium (mmol/L)	1.92	2.21	1.89	2.42	0.035	0.093	<0.001	0.033
P (mmol/L)	1.86	1.77	1.85	1.67	0.102	0.651	0.294	0.713
Ca/P ratio	1.08	1.26	1.12	1.51	0.066	0.115	0.002	0.217
Liver enzymes								
AST (U/L)	147	74.9	450	84.2	68.2	0.126	0.034	0.150
GLDH (U/L)	41.8	9.9	73.6	16.7	14.1	0.240	0.009	0.445
GGT (U/L)	26.5	21.1	58.0	20.3	8.49	0.061	0.011	0.053
Acute phase proteins								
Hp (µg/mL)	2185	1212	1248	353	374	0.106	0.093	0.942
SAA (µg/mL)	126	59.1	132	60.0	22.2	0.904	0.033	0.927

[1] NEFA = non-esterified fatty acids; BHBA = beta-hydroxy-butyrate; BUN, blood urea nitrogen; AST, Aspartataminotransferase; GLDH, Glutamatdehydrogenase; GGT, Gamma-glutamyltransferase; Hp, Haptoglobin; SAA, Serum amyloid A; [2] SEM = standard error of the mean; [3] Effects of lactation number: Lact, cows were grouped in ≤2 lactations ($n = 10$ in LDA group and $n = 11$ in HC group) or >2 lactations ($n = 10$ in LDA group and $n = 9$ in HC group); Status, health status of cows; Int, Lact × Status interaction.

Serum activities of liver enzymes AST, GGT, and GLDH were significantly raised in cows with LDA compared to healthy cows ($p < 0.05$). However, there was no difference regarding the effects of lactation number on these enzymes, except of GGT values being greater in the sick cows with >2 lactations ($p = 0.05$).

Serum concentrations of Hp tended to be greater in cows with LDA ($p = 0.09$), and these cows also displayed elevated SAA concentrations compared to their healthy counterparts ($p < 0.05$). Lactation number and the interaction between this variable and health status showed no effect on both serum APP.

3.2. Milk Production and Composition

Milk parameters measured four weeks after surgery differed between healthy cows and cows with LDA (Table 2). Milk protein tended to be lower whereas the fat/protein ratio and SCC tended to be greater in cows with LDA compared to healthy cows. Additionally, an interaction between lactation number and health status was observed for milk protein content ($p < 0.05$) suggesting that the percentage of milk protein was decreased by LDA in cows >2 lactations, but not in cows with ≤2 lactations. The same interaction was observed for milk fat percentage too. Overall, cows' lactation number, affected several milk parameters, including milk yield, energy-corrected milk, fat and protein. Specifically, cows with >2 lactations had higher daily milk yield and protein yield ($p < 0.001$) than cows with ≤2 lactations. With increasing lactation number, milk fat and protein contents decreased ($p < 0.05$), whereas the daily milk fat yield increased ($p < 0.05$).

Table 2. Milk yield and composition in cows with left abomasal displacement (LDA) or healthy cows (HC) differing in the lactation number.

Milk Variables [1]	≤2 Lactations		>2 Lactations		SEM [2]	*p*-value [3]		
	LDA	HC	LDA	HC		Lact	Status	Int
Milk yield (kg/d)	26.4	27.6	33.4	36.2	1.99	<0.001	0.360	0.708
ECM (kg/d)	27.6	26.1	29.6	34.7	1.96	0.013	0.384	0.114
Fat (%)	4.49	3.60	3.36	3.73	0.22	0.038	0.281	0.010
Protein (%)	3.27	3.22	2.92	3.22	0.08	0.041	0.144	0.043
Fat yield (kg/d)	1.16	0.99	1.10	1.34	0.09	0.146	0.734	0.049
Protein yield (kg/d)	0.85	0.88	0.97	1.16	0.05	0.002	0.067	0.195
Fat/protein ratio	1.37	1.12	1.16	1.15	0.06	0.188	0.062	0.071
SCC (cells/mL)	141.8	79.5	379.8	104.2	66.7	0.173	0.082	0.266

[1] Milk data were recorded 3 times in a week interval starting from 45 d postpartum (almost four weeks after the surgery); ECM, energy-corrected milk = (kg milk × (0.383 × % fat + 0.242 × % protein + 0.7832)/3.1138); SCC, somatic cell count (expressed in × 1000 cells/mL); [2] SEM = standard error of the mean; [3] Effects of lactation number: Lact, cows were grouped in ≤2 lactations (n = 10 in LDA group and n = 11 in HC group) or >2 lactations (n = 10 in LDA group and n = 9 in HC group); Status, health status of cows; Int, Lact × Status interaction.

3.3. Associations between Milk Yield and Serum Variables

Associations between milk yield data and blood variables are shown in Figure 1. Cows with LDA showed a linear increase in serum NEFA,

BHBA, as well as NEFA/cholesterol ratio with increasing milk yield ($R^2 = 0.21$ to 0.45, $p < 0.05$), albeit not showing any significant relationship in healthy cows (Figure 1a–c). Moreover, milk yield was positively linearly related to serum Ca (Figure 1d) as well as Ca:P ratio (data not shown) in healthy cows ($R^2 = 0.35$ to 0.42, $p < 0.01$), whereas an opposite effect was observed for the association between serum Ca and daily milk yield in cows affected by LDA ($R^2 = 0.11$, $p = 0.094$, Figure 1d).

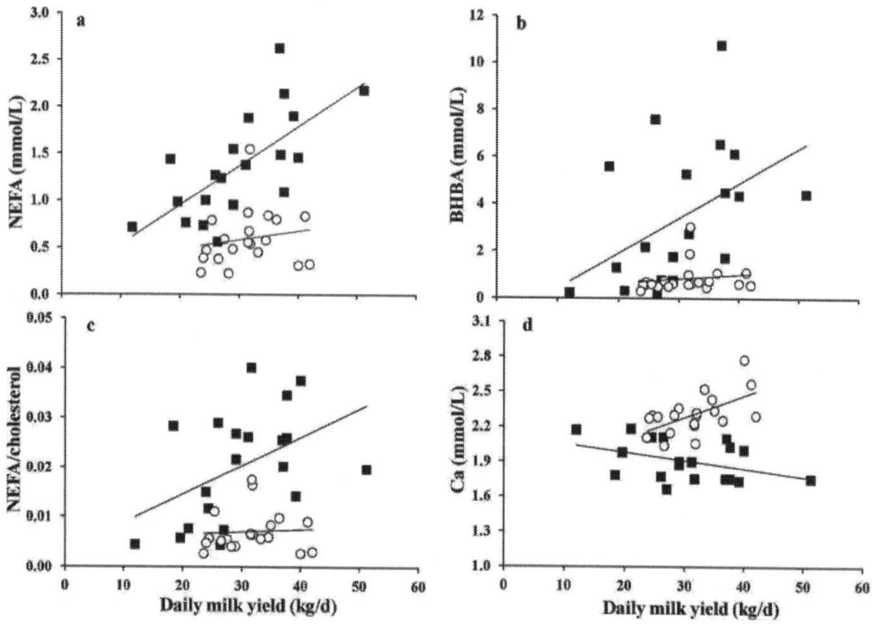

Figure 1. Association between milk yield and serum (**a**) non-esterified fatty acids (NEFA); (**b**) beta-hydroxy-butyrate (BHBA); (**c**) NEFA/cholesterol ratio; and (**d**) calcium in cows affected by left abomasal displacement (■) or not (○). Significant associations in cows affected with left abomasal displacement: NEFA = 0.11 + 0.420 × milk yield, $R^2 = 0.45$, $p < 0.001$; BHBA = −1.06 + 0.149 × milk yield, $R^2 = 0.21$, $p = 0.039$; NEFA: cholesterol = 0.008 + 0.00025 × milk yield, $R^2 = 0.22$, $p = 0.035$; Ca = 2.13 − 0.0072 × milk yield, $R^2 = 0.11$, $p = 0.094$. Significant associations in healthy cows: Ca = 1.72 + 0.018 × milk yield, $R^2 = 0.35$, $p = 0.006$.

4. DISCUSSION

4.1. Metabolic and Inflammatory Responses

This research aimed to investigate metabolic and inflammatory responses associated with LDA in high-producing dairy cows kept under small-scale farm conditions. We observed dramatic increases of NEFA and BHBA concentrations associated with higher SAA in the sick cows, which are typically findings reported for LDA cows [1,5,13]. High concentrations of serum BHBA derive from the oxidation of excessive NEFA in hepatocytes due to an enhanced lipolysis and the lack of energy, which is aggravated in LDA cows because of excessive NEB [1]. A NEB in postparturient dairy cows with LDA has been reported to lead not only to an excessive lipolysis but also to the release of proinflammatory cytokines such as TNF-α from adipocytes, resulting both in elevated serum NEFA and APR conditions with increased hepatic synthesis of SAA [14]. This assumption is supported by our findings of greater NEFA and NEFA/cholesterol, as well as SAA levels in the LDA cows. The elevated ratio of NEFA/cholesterol is indicative of fat infiltration of the liver [23], whereas greater SAA suggest an inflammatory condition in LDA cows [12]. Furthermore, the tendency for higher milk fat/protein ratio and lower protein yield in LDA cows is indicative of a more severe energy deficiency in association with greater body mass losses, which are known to contribute to increasing milk fat/protein ratio [24].

Typically, all cows in the immediate postparturient period experience an energy and nutrient deficit, whereby cows experiencing LDA lower their feed intake long before the occurrence of disease [1]. The latter event aggravates further the energy and nutrient deficit immediately after parturition. This causal relationship likely explains the current findings of dramatic increase of NEFA and BHBA concentrations in the serum of cows with LDA. Several authors claimed serum BHBA concentrations around 1.6 mmol/L in the first week post-partum as BHBA cut-point for prediction of LDA risk [5,25,26]. In the present study, blood samples were collected at the time point of surgery and most of the cows, in particular those with high milk yields were already in a clinical ketotic state, showing about three times higher BHBA values than healthy cows.

In agreement with the present results, elevated serum levels of the enzyme AST have been reported in cows with LDA in previous studies [13,27]. We also observed significant increases in serum GGT and GLDH, which may indicate liver cell damage in LDA cows [13], possibly caused by hepatic lipidosis. In addition, a rise of blood AST activity has been assumed to result from muscle protein mobilization or depletion aiming to deliver glycogenic amino acids as glucose precursors [28], which was supported by the detected lower concentration of milk protein in LDA cows.

Although serum Hp only showed a tendency, the data of SAA support the line of thinking that development of LDA is associated with an inflammatory state. This assumption has been supported by several other reports [12,13,14,15]. The fact that Hp did not significantly differ between groups can likely be explained by the fact that this marker is more involved in the subacute or chronic inflammatory conditions rather than acute inflammation processes in cattle [29]. Despite little evidence linking development of systemic APR to lowering levels of blood Ca [14,15] and also a relationship between low serum Ca and LDA occurrence [15,26,30,31], the mechanisms by which APR is linked to the development of LDA has not yet been established. It is generally accepted that low blood Ca inhibits abomasal motility [9]. Supporting the theory of a disturbed Ca homeostasis, cows with LDA had a decreased serum Ca concentration compared to healthy cows in the present study. The serum concentration of LDA cows was <2 mmol/L, a concentration which has been reported to increase the risk of LDA [31].

4.2. Effects of Lactation Number and Milk Yield on Blood Variables

An important aim of this study was to establish an association between lactation number and milk production with the outcome of metabolic and inflammatory variables. Regarding the effect of lactation number, the present data demonstrated that cows in the third or higher lactation showed higher serum values of NEFA, BHBA, NEFA/cholesterol ratio, and GGT than cows in the first or second lactation. These results led to the assumption that higher lactation numbers exacerbate metabolic alterations associated with LDA likely due to higher milk yields. Indeed, interesting findings of the present study were the association between increasing milk yield and reduced serum Ca, and increasing NEFA, BHBA, and NEFA/cholesterol ratio in LDA cows. The reasons why milk yield only exacerbates metabolic variables, but not inflammatory and liver enzymes in LDA cows, and not in healthy cows, is not known and requires further investigation. It is possible that high milk production leads to greater drainage of nutrients via milk in LDA cows, which cannot be compensated by nutrient intake. It is also likely that high milk production aggravates the catabolic state provoked by LDA, leading to exacerbation of metabolic alterations, which were not observed in healthy cows. It seems that other factors besides milk production are involved in these strong metabolic alterations. Already Martin [32] pointed out that, due to strong metabolic and endocrine changes, parturition could predispose all cows to LDA; however, only certain individuals have other important predisposing factors which favor the development of LDA [33].

A NEB is well-known as the initiating factor in the development of both ketosis and hepatic lipidosis, which have been implicated as risk factors for LDA [8]. The present data indicate that in cows affected by LDA a high daily milk yield results in a deeper or more long-lasting NEB around calving. Oppositely, healthy cows appear to be in less severe NEB, even at comparably higher milk yield levels, which indicates a better adaptation of these cows during the initiation of the lactation.

Interestingly, in healthy cows, serum Ca as well as Ca/P ratio increased with increasing milk yield, whereas the opposite effect was observed for Ca in LDA cows. Despite assumed higher feed intake in healthy cows, this might be related to a higher ability of healthy cows to respond to the developing hypocalcemia by secreting parathyroid hormone, enabling Ca mobilization from bone, as well as enhanced renal reabsorption and efficient intestinal absorption [34]. In contrast, these homeostatic mechanisms appear to fail to operate sufficiently in cows affected by LDA, resulting in hypocalcemic conditions, becoming even more pronounced with increasing milk yield. Thus, the obtained results let assume that increasing milk yield impairs Ca metabolic regulation mechanisms in cows with LDA, whereas this is not the case in healthy cows.

The comparison among groups of cows regarding serum cholesterol should be interpreted with caution, because cows were not matched by lactation day. Serum cholesterol is a variable which is very sensitive to lactation day in cows [2]. Another limitation of this study is the disparity of lactation number of the cows in the LDA and HC group, in a way that eight LDA cows were of four lactations and higher, whereas only two cows in the HC group had this number of lactations. Studies with equal number of lactations between groups are needed to verify the present findings.

5. CONCLUSIONS

Taken together, the study indicated metabolic alterations, liver damage, and inflammation in LDA cows kept under small-scale farm conditions. Our data also suggest exacerbation of metabolic profile and Ca metabolism but not of inflammation and liver health with increasing lactation number and milk yield in LDA cows, and not in healthy cows. From a clinical point of view, the current data suggest the need of a rapid intervention to re-establish the aggravated ketonemia and hypocalcemia, liver health, and inflammation, whereby high-producing LDA cows with >2 lactations require more attention regarding metabolic variables than younger cows.

ACKNOWLEDGMENTS

We acknowledge the assistance of Anita Dockner (Institute of Animal Nutrition and Functional Compounds, University of Veterinary Medicine, Vienna, Austria) with the laboratory analyses. The contribution of Christa Egger-Danner (ZuchtData EDV-Dienstleistungen GmbH, Vienna, Austria) with the recovery of milk and disease data is highly acknowledged. The study was funded by the collaborative project ADDA (ADvancement of Dairying in Austria) of Austrian Research Promotion Agency (FFG) (Grant No. 843543).

AUTHOR CONTRIBUTIONS

Qendrim Zebeli and Leopold Podstatzky-Lichtenstein conceived the study. Qendrim Zebeli and Leopold Podstatzky-Lichtenstein acquired the data. Qendrim Zebeli, Barbara Metzler-Zebeli, Fenja Klevenhusen, Elke Humer, Thomas Wittek analyzed and interpreted the data. Fenja Klevenhusen, Elke Humer, Qendrim Zebeli, Barbara Metzler-Zebeli drafted the manuscript. Qendrim Zebeli, Fenja Klevenhusen, Elke Humer, Barbara Metzler-Zebeli, Thomas Wittek critically revised the manuscript. All authors read and approved the final manuscript.

CONFLICTS OF INTEREST

The authors declare no conflict of interest.

REFERENCES

1. Ospina, P.A.; Nydam, D.V.; Stokol, T.; Overton, T.R. Association between the proportion of sampled transition cows with increased nonesterified fatty acids and beta-hydroxybutyrate and disease incidence, pregnancy rate, and milk production at the herd level. *J. Dairy Sci.* **2010**, *93*, 3595–3601.

2. Gross, J.J.; Kessler, E.C.; Albrecht, C.; Bruckmaier, R.M. Response of the cholesterol metabolism to a negative energy balance in dairy cows depends on the lactational stage. *PLoS ONE* **2015**, *10*.

3. Shaver, R.D. Nutritional risk factors in the etiology of left displaced abomasum in dairy cows: A review. *J. Dairy Sci.***1997**, *80*, 2449–2453.

4. Kelton, D.F.; Lissemore, K.D.; Martin, R.E. Recommendations for recording and calculating the incidence of selected clinical diseases of dairy cattle. *J. Dairy Sci.* **1998**, *81*, 2502–2509.

5. Qu, Y.; Lytle, K.; Traber, M.G.; Bobe, G. Depleted serum vitamin E concentrations precede left displaced abomasum in early-lactation dairy cows. *J. Dairy Sci.* **2013**, *96*, 3012–3022.

6. Guard, D. The costs of common diseases of dairy cattle (proceedings). In Proceedings of the CVC in San Diego, San Diego, CA, USA, 1 October 2008.

7. Raizman, E.A.; Santos, J.E.P. The effect of left displacement of abomasum corrected by toggle-pin suture on lactation, reproduction, and health of Holstein dairy cows. *J. Dairy Sci.* **2002**, *85*, 1157–1164.

8. Cameron, R.E.B.; Dyk, P.B.; Herdt, T.H.; Kaneene, J.B.; Miller, R.; Bucholtz, H.F.; Liesman, J.S.; Vandehaar, M.J.; Emery, R.S. Dry cow diet, management, and energy balance as risk factors for displaced abomasum in high producing dairy herds. *J. Dairy Sci.* **1998**, *81*, 132–139.

9. Doll, K.; Sickinger, M.; Seeger, T. New aspects in the pathogenesis of abomasal displacement. *Vet. J.* **2008**, *181*, 90–96.

10. Koeck, A.; Miglior, F.; Jamrozik, J.; Kelton, D.F.; Schenkel, F.S. Genetic associations of ketosis and displaced abomasum with milk production traits in early first lactation of Canadian Holsteins. *J. Dairy Sci.* **2013**, *96*, 4688–4696.

11. Mömke, S.; Sickinger, M.; Rehage, J.; Doll, K.; Distl, O. Transcription factor binding site polymorphism in the *motilin*gene

associated with left-sided displacement of the abomasum in German Holstein cattle. *PLoS ONE* **2012**, *7*.

12. Ametaj, B.N.; Zebeli, Q.; Iqbal, S. Nutrition, microbiota, and endotoxin-related diseases in dairy cows. *R. Bras. Zootec.* **2010**, *39*, S433–S444.

13. Stengärde, L.; Holtenius, K.; Tråvén, M.; Hultgren, J.; Niskanen, R.; Emanuelson, U. Blood profiles in dairy cows with displaced abomasum. *J. Dairy Sci.* **2010**, *93*, 4691–4699.

14. Guzelbektes, H.; Sen, I.; Ok, M.; Constable, P.D.; Boydak, M.; Coskun, A. Serum amyloid A and haptoglobin concentrations and liver fat percentage in lactating dairy cows with abomasal displacement. *J. Vet. Int. Med.* **2010**, *24*, 213–219.

15. Zebeli, Q.; Sivaraman, S.; Dunn, S.M.; Ametaj, B.N. Intermittent parenteral administration of endotoxin triggers metabolic and immunological alterations typically associated with displaced abomasum and retained placenta in periparturient dairy cows. *J. Dairy Sci.* **2011**, *94*, 4968–4983.

16. Stengärde, L.; Hultgren, J.; Tråvén, M.; Holtenius, K.; Emanuelson, U. Risk factors for displaced abomasum or ketosis in Swedish dairy herds. *J. Dairy Sci.* **2012**, *103*, 280–286.

17. Agger, J.; Alban, L. Welfare in Danish dairy herds 3. Health management and general routines in 1983 and 1994. *Acta Vet. Scand.* **1996**, *37*, 79–97.

18. Markusfeld, O. Periparturient traits in seven high dairy herds. Incidence rates, association with parity, and interrelationships among traits. *J. Dairy Sci.* **1987**, *70*, 158–166.

19. Wolf, V.; Hamann, H.; Scholz, H.; Distl, O. Influences on the occurrence of abomasal displacements in German Holstein cows. *Dtsch. Tierarztl. Wochenschr.* **2001**, *108*, 403–408.

20. Klevenhusen, F.; Hollmann, M.; Podstatzky-Lichtenstein, L.; Krametter-Frötscher, R.; Aschenbach, J.R.; Zebeli, Q. Feeding barley grain-rich diets altered electrophysiological properties and permeability of the ruminal wall in a goat model. *J. Dairy Sci.* **2013**, *96*, 2293–2302.

21. Ertl, P.; Zebeli, Q.; Zollitsch, W.; Knaus, W. Feeding of by-products completely replaced cereals and pulses in dairy cows and enhanced edible feed conversion ratio. *J. Dairy Sci.* **2015**, *98*, 1225–1233.

22. Tempelman, R.J. Experimental design and statistical methods for classical and bioequivalence hypothesis testing with an application to dairy nutrition studies. *J. Anim. Sci.* **2004**, *82*, E162–E172.

23. Kaneene, J.B.; Miller, R.; Herdt, T.H.; Gardiner, J.C. The association of serum nonesterified fatty acids and cholesterol, management and feeding practices with peripartum disease in dairy cows. *Prev. Vet. Med.* **1997**, *31*, 59–72.

24. Geishauser, T.; Leslie, K.; Duffield, T.; Edge, V. An evaluation of protein/fat ratio in first DHI test milk for prediction of subsequent displaced abomasum in dairy cows. *Can. J. Vet. Res.* **1998**, *62*, 144–147.

25. Geishauser, T.; Leslie, K.; Duffield, T.; Edge, V. Evaluation of aspartate aminotransferase activity and ß-hydroxybutyrate concentration in blood as tests for prediction of left displaced abomasum in dairy cows. *Am. J. Vet. Res.* **1997**, *58*, 1216–1220.

26. Seifi, H.A.; LeBlanc, S.J.; Leslie, K.E.; Duffield, T.F. Metabolic predictors of post-partum disease and culling risk in dairy cattle. *Vet. J.* **2011**, *188*, 216–220.

27. Itoh, N.; Koiwa, M.; Hatsugaya, A.; Yokota, H.; Taniyama, H.; Okada, H.; Kudo, K. Comparative analysis of blood chemical values in primary ketosis and abomasal displacement in cows. *J. Vet. Med. A.* **1998**, *45*, 293–298.

28. Herdt, T.H. Ruminant adaption to negative energy balance, influences on the etiology of ketosis and fatty liver. *Vet. Clin. N. Am. Food A.* **2000**, *16*, 215–230.

29. Horadagoda, N.U.; Knox, K.M.G.; Gibbs, H.A.; Reid, S.W.J.; Horadagoda, A.; Edwards, S.E.R.; Eckersall, P.D. Acute phase proteins in cattle: Discrimination between acute and chronic inflammation. *Vet. Rec.* **1999**, *144*, 437–441.

30. Van Winden, S.C.; Jorritsma, R.; Müller, K.E.; Noordhuizen, J.P. Feed intake, milk yield, and metabolic parameters prior to left displaced abomasum in dairy cows. *J. Dairy Sci.* **2003**, *86*, 1465–1471.

31. Van Saun, R.J.; Todd, A.; Varga, G.A. Serum mineral concentrations and risk of periparturient disease. *Proc. Am. Ass. Bov. Pract.* **2005**, *38*, 178–179.

32. Martin, W. Left abomasal displacement: An epidemiological study. *Can. Vet. J.* **1972**, *13*, 61–68.

33. Wittek, T.; Sen, I.; Constable, P.D. Changes in abdominal dimensions during late gestation and early lactation in Holstein Friesian heifers and cows and their relationship to left displaced abomasum. *Vet. Rec.* **2007**, *161*, 155–161.

34. Goff, J.P. Major advances in our understanding of nutritional influences on bovine health. *J. Dairy Sci.* **2006**, *89*, 1292–1301.

CHAPTER 3

Characteristics of Lactation Curves of the Kenya Alpine Dairy Goats in Smallholder Farms

Andrew Gitahi Marete[1*], Reuben Oyoo Mosi[2], Joshua Oluoch Amimo[1,3], Joseph Owino Jung'a[1]

[1]Department of Animal Production, University of Nairobi, Nairobi, Kenya

[2]School of Agriculture, Food Security and Biodiversity, Bondo University College, Bondo, Kenya

[3]Department of Food Animal Health Research Program, The Ohio State University, Columbus, USA

ABSTRACT

Lactation curves are a graphical representation of the milk production profile of a doe from parturition to drying up. Their shape provides information about the productivity of the doe and offers a means of explaining features of the milk production patterns of each animal. A total of 2732 daily morning milk records from 610 does of the Kenyan Alpine dairy goats' genetic groups (50% Alpine, 75% Alpine, 87.5% Alpine and > 87.5% Alpine) and local goats (0% Alpine) kept in smallholder farms were used to evaluate factors affecting milk yield and to examine the characteristics of their lactation curve. A nonlinear mixed model was used to fit the lactation curves to all does simultaneously. The Wood's (1967) equation was fitted within each genetic group and parity to generate genetic group and parity lactation curves. The mean lactation period was 218 ± 46 days and the model accounted for 88% of the total variation. Significant differences ($P < 0.05$) between genetic groups were observed in their lactation curve parameters. The estimated week of peak milk yield post kidding was: 2, 4, 6, 6 and 8 weeks; and peak yield was 0.32, 0.75, 0.91, 0.99 and 1.02 Kg/day, for 0% Alpine, 50% Alpine, 75% Alpine, 87.5% Alpine and >87.5% Alpine genetic groups respectively. Genetic group did not significantly affect rate of increase to peak yield ($P > 0.05$) and rate of decline from peak ($P > 0.05$) or persistency ($P > 0.05$). Parity significantly affected rate of increase to peak, rate of decrease from peak and persistency ($P < 0.01$). The month of kidding significantly affected the rate of increase to peak ($P < 0.05$) and persistency, but not rate of decrease from peak. The synchronization of breeding with season has a practical implication for the maximization of lactation yield when considered in combination with other biological and economic constraints. The superior production of the pedigree animals supports the development of composite breed types in Kenya to take advantage of the fitness of indigenous breeds, the productivity of imported dairy breeds, heterosis, and the potential for selection within the composite to improve productivity in later generations.

KEYWORDS

Kenya Alpine Dairy Goat; Lactation Curve; Smallholder Farms

1. INTRODUCTION

The term "lactation curve" is used to refer to the curve representing the rate of milk yield with advance in lactation. Goats can lactate from kidding to kidding; the challenge would therefore be to estimate the lactation curve which would indicate the days in milk (DIM) up to peak during the lactation period. Woods model has been used in majority of studies to characterize parameters of lactation curves since it had a residual low fluctuation around zero ([1] -[3] reported characteristics of goats lactation curve (maximum yield and time of maximum yield) separately by parity groups (first, second and third or greater)). [4] analyzed performance data of goats raised in a stall-fed system in northern Mexico in order to compare high grade goats with low grade crosses of Alpine, Granadina, Nubian, Saanen and Toggenburg sires with local Mexican goats. Alpine, Saanen and Toggenburg crosses had greater milk production, lactation length and efficiency (milk production/post-kidding body weight) than Granadina and Nubian crosses or local does. Average increase for milk production, lactation length and efficiency over local goats for low-grade Alpine was reported to be 46%. Effects of year of kidding, season of kidding and age were reported to be most significant with goats having two or more kids at birth having greater milk production efficiency and body weight than goats with one kid. As reported by [5] , daily milk yield increased rapidly from kidding to a peak in 35 days, followed by a gradual decline until milking was no longer practical. An abrupt decline in milk yield after the peak increases production costs because yield was distributed less equally over the complete lactation [5] .

[6] characterized lactation curves among crosses of Galla and East African with Toggenburg and Anglo Nubian goats and found that Wood's equation fitted well the lactation curves of the different breed groups with R^2 values ranging from 85% to 92%. There were differences in lactation curve parameters among the genetic groups. The four-way cross had the highest average peak morning milk yield (1.116 kg/d) and estimated total morning milk yield (112.582 kg).

In developing countries, the performance of high-yielding breeds imported from countries with highly advanced production systems is often negatively affected due to genotype-environment interactions ([7] -[10]). This study provides the basic framework needed for the future development of computationally feasible techniques to evaluate the dairy potential of a Kenya Alpine dairy goat given the initial milk production and a future time, t. The aim of this study was to use an appropriate mathematical model to obtain parameter estimates that would allow a better understanding of the biological nature of the different shapes of lactation curves of dairy goats in Kenya.

2. MATERIALS AND METHODS

2.1. Animals

The Kenya Alpine dairy goat was used in this study. The breed was developed by crossing and grading-up of the East African goat using German Alpine buck in an effort to providing alternative source of income to smallholder farmers' in Kenya and to improve nutrition at household level. For over 10 years now, Kenya Alpine dairy goats were bred through natural service, a process characterized by buck rotation facilitated by the Dairy Goat Association of Kenya (DGAK). The goats were usually stall-fed and every household kept between 3 - 6 goats on average. Milking was usually done in the morning after kids have suckled. The goats were housed and fed in a shed. Currently there are four genetic groups of the Kenya Alpine dairy goats and include:

1) 50% Alpine-F_1: Pedigreee Alpine Buck (PAB) × Local doe.

2) 75% Alpine-Back cross 1 (R_1): PAB × F_1 Females.

3) 87.5% Alpine-Back cross 2 (R_2): PAB × R_1 Females.

4) >87.5% Alpine:

a) Interse mating of Back cross 2: R_2 Male × R_2 Female.Offspring will remain 87.5% or more.

b) PAB × R_2 Female—Grading-up. Offspring will be 93.25% or more.

There has been stabilization of the genetic group with >87.5% alpine genes through breeding, intense selection and culling to form the Kenya Alpine goat breed. Kenya Alpine dairy goats are usually found in the Central and Eastern highlands of Nyeri and Meru districts respectively. Central province was located at 0°45'S 37°0'E and in the study period (1999-2009), this area's districts experienced a closely related climatic pattern with a daily temperature range of 18.0°C ± 5.1°C and an annual rainfall ranging from between 607.8 mm to 1701.0 mm per annum. Generally, the long rains were between March to May, while the short rains were from October to December. There were two dry periods which occurred between June to September and January to February. Eastern Province was located at 0°0'N 38°0'E and areas around Mt. Kenya experience cold climate all year round due to its high altitude. Meru Central was in the humid zone of Eastern while Embu district was in the sub-humid zone. The annual rainfall ranges from 701.0 mm to 1800.0 mm with a mean diurnal temperature range of 19.5°C to 24.5°C. The vegetation was mainly forested land in parts of Meru to agricultural lands in Embu district.

In recent years, the Kenya Alpine goats have spread to lower potential areas (Thika and Murang'a Districts) compared to the original entry areas. These areas are characterized by a humid to sub-humid climate with long, wet and cold wet seasons. Though they are zero-grazed, the kidding season is usually not synchronized but rather depends on conditions of pasture.

2.2. Data

A total of 610 lactating goats were examined over a period of five lactations. As presented in Table 1, 2732 daily morning milk records from the 610 lactating goats were used to fit each individual curve because the does usually nurse their kids in the late afternoon. The age at first kidding (AFK) in days, was calculated as the difference between the date of first kidding (DoK) and the doe's date of birth (DoB). The age at first service (AFS) in days, was estimated as the difference between the date of first service (DoS) and the doe's date of birth (DoB). The open days (OD) was estimated as the difference between the date of second service (DoS2) and the date of first kidding (AFK) i.e. $OD_n = DoS_{n+1} - AFK_n$ where n was the parity.

2.3. Statistical Analysis

Lactation records made by 610 goats over a period of five lactations were used to characterize the lactation curve. Two analyses were conducted: The first analysis was to estimate lactation curves of each genetic group represented in the data using Wood's (1967) equation. Morning milk yields during each week of lactation within each genetic group were used to fit genetic group lactation curves separately using the gamma function:

$$Y_t = at^b e^{-ct}$$

where Y_t, is the average morning yield (kg) in week t of lactation; "a" is a scaling factor representing initial average morning yield; "b" is the rate of increase to peak production; "c" is the rate of decline after peak production; and "e" is the exponential. The model parameters were estimated by nonlinear least squares using PROC NLIN of SAS (SAS, 2002).

Table 1. Distribution of lactation records by genetic group and parity.

	Parity					
Genetic group	**1**	**2**	**3**	**4**	**5**	**Total**
0% Alpine (local)	42	22	20	19	21	**124**
50% Alpine (foundation)	312	77	54	38	37	**518**
75% Alpine (intermediate)	549	123	83	54	52	**861**
87.5% Alpine (appendix)	550	125	84	55	53	**867**
>87.5% Alpine (pedigree)	66	116	78	52	50	**362**
Total	**1519**	**463**	**319**	**218**	**213**	**2732**

Peak yield was defined as: $a(b/c)^{b}e^{-b}$ and time of peak yield was: (b/c)

The log transformed gamma function was used to estimate persistency (s) as:

$$s = -(b+1)\ln(c)$$

The second analysis involved fitting Wood's equation results for each genetic group to obtain individual status curve parameters. A total of 2732 daily morning milk records were used to fit individual curves because the does were allowed to nurse their kids in the late afternoon. Different factors affecting lactation curve parameters within the genetic groups and parities were assessed. The statistical model used is as shown in model I below:

Model I: $Y_{ijklm} = \mu + YR_i + S_j + B_k + V_1 A_{ijkl} + V_2 A_{ijkl}^2 + e_{ijklm}$

where, Y_{iiklm} represents the model parameter a, b, c, or persistency defined as s = −(b + 1)ln(c) of doe 1, of breed type k, that kidded in season j, in year i

μ is the underlying constant in each observation;

YR_i is year of kidding (1999-2009);

S_j is season of kidding (wet = April, May, June, October, and November; cold dry = July, August, and September; hot dry = December, January, February, and March);

B_k is the genetic group of doe; and V_1 and V_2 are linear and quadratic partial regression coefficients, respectively, on age of dam (A_{ijkl}) at kidding.

e_{ijklm} is the effect of random effects, assumed to be normally distributed with a mean 0 and a variance $\sigma^2 = 1$

3. RESULTS

3.1. Lactation Curve Parameter Estimates by Genetic Group

Lactation curve parameter estimates for each genetic group are presented in Table 2. Wood's equation fitted well the lactation curves of the different genetic groups with R^2 values ranging from 85% to 91%. There were differences in lactation curve parameters among the genetic groups. Genetic group with >87.5% Alpine genes had the highest average peak morning milk yield (0.813 ± 0.045 kg/d) and estimated total morning milk yield (108.8 kg @ 120 d). Peak milk yield for genetic group with 50% Alpine was at week 3 of lactation which was earlier than for 87.5% Alpine and 75% Alpine genetic group which peaked in week 4 of lactation.

Figure 1 shows the estimated lactation curves for each of the genetic groups. The lactation curves were consistent with earlier findings that goats have flatter and more persistent curves than cattle ([11]). Persistency of lactation yield was measured from the log transformed gamma function which explained 89% of variation in daily yields, similar to that reported by [6] ; effects of month of kidding ($P < 0.05$) and parity ($P < 0.0001$), were significant for the various lactation curve traits. Peak and lactation yields were lower, but persistency was higher during first lactation.

3.2. Lactation Curve Parameter Estimates by Parity

The effect of parity number on lactation curve parameters "a" and "b" was highly significant ($P < 0.001$ and $P < 0.01$ respectively) whilst parity had no significant effect on parameter "c". Does which kid for the first time had the lowest initial milk yields, which increased gradually till fourth lactation and then decreased thereafter (Figure 2). The analysis of the lactation curve showed the highest value of "a", was observed in the third lactation and it was significantly ($P < 0.01$) different from all other lactations except those in the fourth parity (Table 3). The highest value for "b" was observed in the second and fifth parities and was significantly ($P < 0.05$) different from the third parity. The value for "c" was lowest in the third parity but was not significantly different from any other parity.

Table 2. Lactation curve parameters by genetic group.

Genetic group	No. of Does	Lac records	*a* (kg)	*b*	*c*	*s*	Time to peak yield (wk)	peak day yield(kg)	120d yield (kg)	R^2 (%)
0% Alpine	53	124	$0.345 \pm 0.028^{*}$	$0.149 \pm 0.15^{*}$	$0.082 \pm 0.037^{*}$	2.87^{ns}	2	0.3249	38.989	85
50% Alpine	93	518	$0.733 \pm 0.028^{*}$	$0.057 \pm 0.04^{*}$	0.014 ± 0.007^{ns}	4.51^{*}	4	0.7501	90.009	91
75% Alpine	157	861	$0.721 \pm 0.040^{*}$	$0.292 \pm 0.06^{*}$	0.049 ± 0.009^{ns}	3.90^{ns}	6	0.9067	108.807	86
87.5% Alpine	137	867	$0.813 \pm 0.022^{*}$	$0.248 \pm 0.03^{*}$	$0.042 \pm 0.004^{*}$	3.96^{**}	6	0.9855	118.256	89
>87.5% Alpine	170	362	$0.813 \pm 0.045^{*}$	$0.216 \pm 0.06^{*}$	$0.028 \pm 0.009^{*}$	4.35^{ns}	8	1.0184	122.214	90

^{ns} = not significant ($P > 0.05$), $^{*}P < 0.05$, $^{**}P < 0.001$.

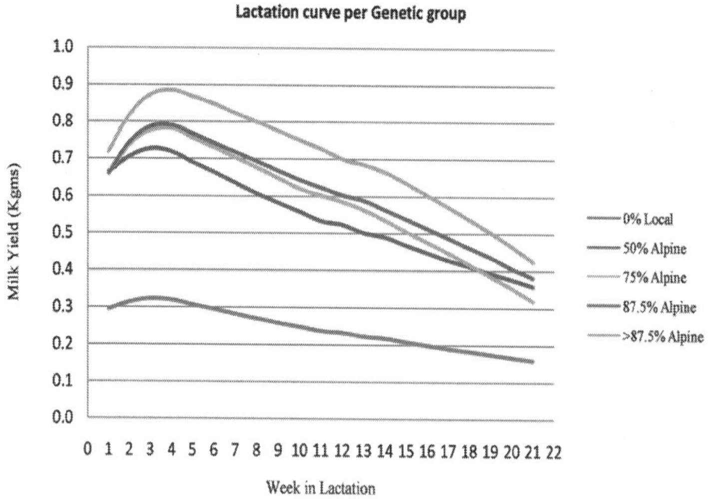

Figure 1. Lactation curve by genetic group.

Parity had a significant effect on lactation parameters as shown in Table 3 above. Higher parity animals had higher intercept values; a greater incline parameter (b) while the effect on the declining parameter (c) was not significant in parity 3 and parity 5 (P > 0.05). [12] concluded that the milk secretory tissue in primiparous animals takes longer to reach its peak activity. In the present study, third and fourth parity goats had a longer duration to peak production compared to animals in first and second parity. Goats in the third parity had the highest initial milk yield but lowest rate of increase up to the peak milk yield. This result can be related to the higher milk yield of animals in this group. [13] reported a progressively greater incline between kidding and peak milk production for milk yield with increasing parity number. Similar results for the lactation curve parameter were reported by [6] [14] -[16] .

3.3. Factors Affecting Lactation Curve Parameters

Genetic group of the animal did not have a significant effect on rate of increase to peak yield (P > 0.05), rate of decline from peak (P > 0.001) and persistency (P > 0.001). The lactation number (parity) had significant effect on rate of increase to peak, rate of decline from peak and persistency (P < 0.01) maybe related to development and maturity of mammary glands and acclimatization to tropical environment.

As presented in Table 4, the month of Kidding (an indication that there is seasonal variation) had a significant effect on the rate of milk increase to peak ($P < 0.05$) and persistency; however it did not have any significant effect from the rate of decrease from peak. The month of kidding and parity have a significant effect in milk production, ($P < 0.05$). The area of birth, the age at first service and the year of kidding do not have any significant effect on the lactation, ($P > 0.05$).

Goats kidding between November to December produced the highest milk yield followed by those that kid between January to March and finally those that kid between May to July. [6] reported different shapes of lactation for season of kidding which may be due to constrains in feed availability. [17] and [18] concluded that a clear separation cannot be found among lactation curves pertaining to different seasons of birth. The effect of season of kidding is largely dependent on the availability of green fodder and management practices adopted to protect the animals against adverse climatic conditions. The cropping pattern in Eastern and central parts of Kenya highlands is such that during certain parts of the year e.g. December to March, abundant green fodder is available but there is an acute shortage from April to June due to dry and hot weather. Goats kidding in May were probably in higher body condition at the time of kidding, had the advantage of abundant green fodder in early lactation and therefore rapidly rose to a high peak yield.

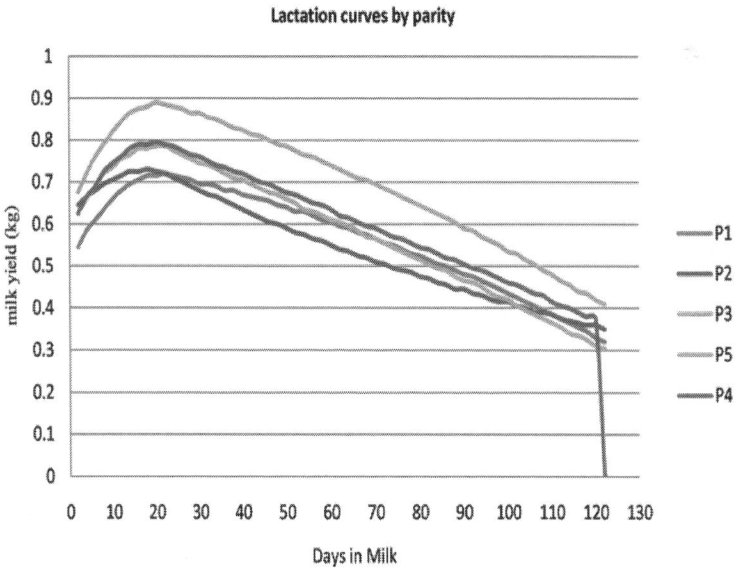

Figure 2. Pooled lactation curves by parity.

Table 3. Lactation curve parameters by parity.

Parity	No. of does	Lactation records	*a* (kg)	*b*	*c*	*s*	R^2 (%)
1	556	1519	$6.08 \pm 0.461^{*}$	$0.10 \pm 0.02^{*}$	$0.040 \pm 0.001^{*}$	3.55^{*}	78
2	353	463	$6.49 \pm 0.404^{*}$	$0.12 \pm 0.01^{*}$	$0.044 \pm 0.001^{*}$	3.50^{*}	79
3	226	319	$7.45 \pm 0.356^{**}$	0.10 ± 0.01^{ns}	0.032 ± 0.001^{ns}	3.78^{**}	89
4	144	218	$7.21 \pm 0.328^{**}$	$0.11 \pm 0.01^{**}$	$0.036 \pm 0.001^{**}$	3.70^{**}	81
5	90	213	6.80 ± 0.337^{ns}	$0.11 \pm 0.01^{*}$	0.032 ± 0.001^{ns}	3.82^{ns}	84

^{ns} = not significant ($P > 0.05$), $^{*}P < 0.05$, $^{**}P < 0.01$.

Table 4. Mean squares of variables from ANOVA of lactation curve traits for Kenya Alpine dairy goats.

Source of variation	DF	*a*	*b*	*c*	*s*
Genetic group	3	12.53 ± 5.035^{ns}	$0.032 \pm 0.0152^{*}$	$0.005 \pm 0.0030^{**}$	6.00^{**}
Type of birth (ToB)	2	$7.03 \pm 0.187^{***}$	0.125 ± 0.099^{ns}	$0.231 \pm 0.0024^{**}$	5.43^{**}
Year of kidding (YoK)	9	$6.442 \pm 0.5516^{*}$	$0.318 \pm 0.042^{*}$	$0.036 \pm 0.098^{**}$	8.94^{ns}
Month of kidding (MoK)	11	$6.77 \pm 1.007^{*}$	$0.340 \pm 0.0242^{**}$	0.111 ± 0.981^{ns}	3.05^{*}
Area of birth (AoB)	4	7.538 ± 0.233^{ns}	1.043 ± 1.992^{ns}	0.287 ± 1.03^{ns}	3.63^{**}
Parity	4	$6.806 \pm 0.940^{**}$	$0.15 \pm 0.032^{**}$	$0.037 \pm 0.010^{**}$	3.79^{**}
Age at first service (AFS)	-	5.976 ± 0.4002^{ns}	$0.385 \pm 0.0151^{*}$	0.049 ± 0.007^{ns}	7.89^{ns}

^{ns} = not significant ($P > 0.05$), $^{*}P < 0.05$; $^{**}P < 0.01$, $^{***}P < 0.001$.

4. MILK YIELD

The outstanding efficiency of the Wood model in this study further emphasizes why the model has been used in most lactation curve studies. [6] used the model to provide a good fit to goat lactation data in Galla, East African and Toggenburg goats. It was also used to understand the main factors that influence production levels and general forms of lactation curves ([19] [20]). Wood's equation fitted well the lactation curves of the different genetic groups with R^2 values ranging from 85% to 91%. The Kenya Alpine goat averaged at least twice as much milk yield as the indigenous breeds for both peak and total yields. The peak yield observed in this study agrees with earlier reports on tropical goats ([11] [21] [22]). Also, the observed peak day conforms to the range reported for tropical goats ([6]) and sheep ([23] [24]). The observed total yield agrees with an earlier report on the same breed of goat ([22]) but it was higher than those of Adal goats ([25]), East African and Galla goats ([6]) and Nigerian Sahel and West African Dwarf goats ([26]). Though all curve parameters differed with respect to genetic group of doe, year, and age of doe, season of kidding significantly affected the rate of decline from peak (parameter c). This was due to changes in pasture availability over the seasons, consistent with that reported by [18] . Since season effects are related to the amounts of rainfall and hence the availability of pasture, these effects could be explained in terms of body condition of the animals if and when supplementations were started. Similar to findings by [6] on East African and Galla goats raised under similar environment, supplementation played a role in observed higher 120-day milk yields.

The persistency over the 120-day lactation period in the Kenya Alpine dairy goats was high (14.38 days) and signifies the ability of these does to maintain their milk production throughout the lactation period. This is far higher than the ones reported for East African goats and their crosses (2.87 to 4.50 days) with the same 120-day lactation period ([6]). The strong negative correlation between parameters "a" and "c" suggests that Kenya Alpine dairy goats selected for high initial milk yield may be highly persistent in milk production. This is because lactation persistency is negatively correlated with rate of decline. A similar trend to that observed in the lactation curve characteristics of Kenya Alpine dairy goats has been reported by other workers on goats ([27]) and sheep ([24] [28]). The described pattern also agrees with reports that lactation curves of goats are flatter and more persistent than those of cattle ([6] [29]). Two reasons, management and within-herd improvement, may be advanced for the effect of herd on the shape of lactation curves in Kenya Alpine dairy goats. Although the zerograzed Alpine herd is generally characterized by stall feeding on a daily basis, there are between herd differences in the nature and timing of food supplementation to the animals. Also, the Kenya Alpine dairy goat farmers practice some level of selection based on the individual farmers' acquired knowledge through personal experience; thus leading to differences in productivity between the herds.

Effects of season of kidding on lactation curve parameters and shape in goats have also been reported in other studies ([6] [30]) and suggest that strategic seasonal breeding is important to maximize milk yield per doe ([6]). The high persistence of the early dry season lactation may be explained by the fact that the declining phase of milk production coincided with the period of low or minimal nutritional stress due to the onset of availability of varieties of crop residues resulting from crop harvest.

According to Table 5, seasonal effect (reflected by the month of kidding) was significant and this contributed to the criss-crossing pattern as seen in Figure 2. [20] obtained a criss-crossing pattern of lactation curves when the parameters describing the ascending or descending phases were significantly influenced by any given factor of consideration due to different curve shapes. They also observed that factors affecting only the parameter describing the level of production resulted in curves having similar shapes but on different levels. Similarly, parity effects have been reported for lactation curves of East African goats by the same authors. In this study, parity 5 had the highest level of production, followed by parties 3 and 4. The lowest production level was for does of parity 1. Parities l, 2 and 3 had their production peaks at week 3.5 while parities 4 and 5 had theirs at week 2. All these point to the fact that it may not be necessary to cull does purely on the basis of performance before parity 5. However, preliminary culling for herd replacement can make use of lactation performance at parity 3 in Kenya Alpine dairy goats. In the ANOVA calculations (Table 5), the findings of the non-significant effects of parity could be ascribed to the processes of accelerated cell proliferation of the milk producing udder tissues, which have been shown to influence the shape of the lactation curves. This was because at the beginning of lactation of the first parity, only a small proportion of milk secretion tissues had developed which might have been too few to result into significant differences between the parity ([31]).

Although found non-significant in this study, higher litter size gave rise to a lower level of milk production. This differed from most other reports on lactating goats ([32] -[34]), but was in agreement with the reports of [35] and [20] support the present observed trend. The curves developed from the gamma-type model as influenced by the various factors encountered in the present study, suggest that the nature of the curves could provide a basis for planning and adjustment in the management of herds, particularly with regards to culling and assessment of the nutritional and health status of animals. These findings may also provide a basis for developing unbiased methods of comparison among animals with incomplete lactation records for genetic evaluation purposes ([36]).

Table 5. Analysis of variance for the parameters of lactation milk yield.

Source	DF	Sum of squares	Mean square	F-value	Pr > F
Genetic group	3	7959.1	2653.02	1.26	0.2914
Type of birth (ToB)	2	2590.3	1295.16	0.62	0.5419
Year of kidding (YoK)	9	19,007	2111.84	1.01	0.4411
Month of kidding (MoK)	11	45,319	4119.94	1.96	0.0405
Area of birth (AoB)	4	1733	433.25	0.21	0.9344
Parity	4	74,780	18694.92	8.9	<0.0001
Age at first service (AFS)	-	661,030	2532.68	1.21	0.1416
R^2	**CV (%)**	**RMSE**	**µDIM**		
0.88	21.05	45.83	217.69		

CV: coefficient of variation; DIM: days in milk; RMSE: root mean square error.

5. MILK YIELD

The different genetic groups had different milk yields. Goats in the pedigree genetic group (>87.5% pure) produced the most amount of milk at peak week with a yield of 1.0184 kg/d with a 120 day total milk yield of 122.2 kg. Compared to previous studies by [6] , cross-breds were shown to have higher peak morning milk yield of 1.116 kg/d and estimated total morning milk yield of 112.58 kg at 122 days. The high milk production of the pedigree was attributed to high contribution of milk production genes of German Alpine origin during grading-up. Appendix performed similar to Pedigree and took 6 weeks to reach peak yield and had a daily peak yield of 0.9855 kg/d with total peak morning milk yield of 118.3 kg at 120 days. This was because of the two classes, none was given preference in terms of feed supplementation hence the only limiting factor was genetic.

Seasonality in milk production was not evident although there were fluctuations in the trait over the study period. It must be recalled that milking was carried out after the kids had suckled hence the total milk production per lactation would be much higher. The smallholder environment did not thus adversely affect milk productivity. The yields obtained in this study were comparable to those observed by [14] for exotic dairy goats under research station conditions in Kenya, but were higher than those reported by [16] for the same breed raised in tropical environments. An over-emphasis on pedigree information and actual weight of an animal at sale as the main determinants of the value placed on an animal by the farmers' association led to most of the smallholder farmers not keeping records on individual animal milk production, but rather they tended to pool together information on total milk sold. Despite the Kenya Alpine goat population having expanded greatly in the region, production and fertility records were few. The value of keeping individual animal records to the farmers could only be evident if the records centrally collected were processed and analyzed, with feedback on progress achieved thereafter being given to the farmers. This would need to be addressed for improved productivity of the Kenya Alpine on smallholder farms.

For all genetic groups, milk yield increased gradually from first parity to fifth similar to previous studies by [15] and [16] , where high lactation yields increased with parity, and since all the genetic group were in similar environments with relatively high rainfall, the comfort zones of the Alpines was surpassed.

6. CONCLUSION

The results demonstrate that the Kenya Alpine goat can perform and thrive reasonably well under low-input farmers' conditions in Kenya. The goats were able to adapt, grow well and produced reasonably high quantities of milk when fed locally available feed resources. Wood's model for the lactation curve in cattle closely fitted lactation curves of the goats. Lactation curve parameters differed by genetic group, season of kidding and age of dam at kidding. >87.5% Alpine genetic group had the highest peak yield and produced the greatest 17 wk total lactation yield. Indigenous breeds peaked in week 2 of lactation and were the least persistent. The superior production of the >87.5% Alpine genetic group supports the development of composite breed types in Kenya to take advantage of the fitness of indigenous breeds, the productivity of imported breeds. Further improvement after the composite breed is stabilised can be based on selection. Significant seasonal effects may have practical implications in determining optimal season of breeding to maximize total lactation yield. However, a high demand for the perceived good quality animal from the resource poor farmers negatively influences any possible selection for improved milk productivity as sale of live animals fetches premium prices. The goat "breeders" need to be encouraged to keep records on individual animal's milk production, and price of animals should be based on potential milk production rather than registration status, supported by regular genetic evaluation of the animals with feedback to the producers on the important traits. A successful recording system must be based not only on the selection goal (genetics has long-term results, especially in harsh conditions), but must also foresee extension services as a means to send a technician to the herd and to give advice to the farmers ([37]). Future developments should focus on the deliverance of an affordable and practical scheme aimed at the requirements of both small scale and commercial goat producers hence the registration of the different classes needs to be vetted at different levels, to avoid erroneous registration of animals in classes that they do not belong.

ACKNOWLEDGEMENTS

The authors wish to acknowledge the German International Development (GIZ) Promotion of Private sector development in Agriculture (PSDA) for funding this study. We also acknowledge the Dairy Goat Association of Kenya (DGAK) for support and availing data collected through their association; the smallholder farmers in Nyeri, Murang'a, Embu, Kirinyaga and Vihiga for availing data on their animals, and the Department of Animal production in the University of Nairobi for the provision of facilities to carry out the study and data analysis.

REFERENCES

1. Wood, P.D.P. (1967) Algebraic Models of the Lactation Curves for Milk, Fat and Protein Production with Estimates of Seasonal Variation. Animal Production, 22, 35.

2. Falagán, A., González, C., Pérez, S.J., Goicoechea, A. and Romero, C. (1991) Composition and Production Curve in the Goat's Milk. Chemie, Mikrobiologie, Technologie der Lebensmittel, 13, 76-82.

3. Fernández, C., Sánchez, A. and Garcés, A. (2002) Modelling the Lactation Curve for Test-Day Milk Yield in Murciano-Granadina Goats. Small Ruminant Research, 46, 29-41.

4. Montaldo, H., Almanza, A. and Juiirez, A. (1995) Genotype, Age and Season Effects on Lactation Curve Shape in Goats. Journal Small Ruminant Research, 24, 195-202.

5. Muhammad, A., Philip, J., Peter, R., Muhammad, S., Muhammad, A. and Maroon, E. (2009) Factors Affecting the Shape of the Lactation Curve in Nile-Ravi Buffaloes in Pakistan. Pakistan Journal of Zoology, 9, 201-220.

6. Ruvuna, F., Kogi, J.K., Mkuu, S.M. and Taylor J.F. (1995) Lactation Curves among Crosses of Galla and East African with Toggenburg and Anglo Nubian Goat. Small Ruminant Research, 16, 1-6.

7. Bondoc, O.C., Smith, C. and Gibson, J. (1989) A Review of Breeding Strategies for Genetic Improvement of Dairy Cattle in Developing Countries. Animal Breeding, 57, 819-829.

8. Rege, J.E.O. (1991) Genetic Analysis of Reproduction and Productive Performance of Cattle in Kenya. Journal of Animal Breeding and Genetics, 108, 412-423.

9. Smith, L.A., Cassell, B. and Pearson, R.E. (1988) The Effect of Inbreeding on the Lifetime Performance of Dairy Cattle. Journal of Dairy Science, 81, 2729-2737.

10. Ojango, J.M.K., Ducrocq, V. and Pollott, G.E (2005) Short Communication Survival Analysis of Factors Affecting Culling Early in the Productive Life of Holstein-Friesian Cattle in Kenya.

11. Devendra, C. and McLeroy, G.B. (1982) Goat and Sheep Production in the Tropics. Longman, Harlow, Essex, UK.

12. Akpa, G.N., Asiribo, O., Oni, O.O. and Alawa, J.P. (2001) The Influence of Non-Genetic Factors on the Shape of Lactation Curves in Red Sokoto Goats. Animal Science, 72, 233-239.

13. Gipson, T.A. and Grossman, M. (1990) Lactation Curves in Dairy Goats: A Review. Small Ruminant Research, 3, 383.

14. Odenya, W.O. (1984) Milk Production Yields from Toggenburg and Black Bedouin Goats under Medium Rainfall Zone. Ministry of Livestock Development, Kenya: Sheep and Goats Development Project, Technical Note No. 21, 5.

15. Andonov, S., Ødegård, J., Boman, I.A., Svendsen, M., Holme, I.J., Adnøy, T., et al. (2007) Validation of Test-Day Models for Genetic Evaluation of Dairy Goats in Norway. Journal of Dairy Science, 90, 4863-4871.

16. Gipson, T.A. (2002) Preliminary Observations: Inbreeding in Dairy Goats and Its Effects on Milk Production. In: Proceedings of 17th Ann. Goat Field Day, Langston, 51-56.

17. Tekerli, M., Akinci, Z., Dogan, I. and Akcan, A. (2000) Factors Affecting the Shape of Lactation Curves of Holstein Cows from the Balikesir Province of Turkey. Journal of Dairy Science, 83, 1381-1386.

18. Catillo, G., Macciotta, N.P.P., Carretta, A. and Cappio-Borlino, A. (2002) Effects of Age and Calving Season on Lactation Curves of Milk Production Traits in Italian Water Buffaloes. Journal of Dairy Science, 85, 1298-1306.

19. Wahome, R.G. (1987) Milk Production Potential of Free Ranging Small East African Goats. Thesis, University of Nairobi, Nairobi.

20. Wahome, R.G., Carles, A.B. and Schewartz, H.J. (1994) An Analysis of Variation of Lactation Curves of Small East African Goats. Small Ruminant Research, 15, 1-7.

21. Morand-Fehr, P. and Sauvant, D. (1978) Nutrition and Optimum Performance of Dairy Goats. Livestock Production Science, 5, 203-212.

22. Osinowo, O.A. and Abubakar, B.Y. (1989) Appropriate Breeding Strategies for Small Ruminant Production in West and Central Africa. In: Adeniji, K.O., Ed., Proceedings of the Workshop on the Improvement of Small Ruminants in West and Central Africa, Ibadan, 21-25 November 1988, 71-84.

23. Lloyd, D.H. (1963) The Milk Production of Merino Ewes at Pasture. Australian Journal of Agricultural Research, 14, 824-838.

24. Hassan, H.A. (1995) Effect of Crossing and Environmental Factors on Production and Some Constituents of Milk in Ossimi and Saidi Sheep and Their Crosses with Chios. Small Ruminant Research, 18, 165-172.

25. Awgichew, K., Yacob, Y. and Fletcher, I. (1991) Productivity of Pure-Bred Adal and Quarter Bred Saanen 5 Adal goats in Ethiopia. Dairy Science Abstracts, 53, 4951.

26. Egwu, G.O., Onyeyili, P.A., Chibuzo, G.A. and Ameh, J.A. (1995) Improved Productivity of Goats and Utilization of Goat Milk in Nigeria. Small Ruminant Research, 16, 195-201.

27. Mukundan, G. and Bhat, P.N. (1983) Lactation Curve in Malabari Goats and Their Saanen Halfbreds. Indian Journal of Animal Science, 53, 666-669.

28. Sakul, H. and Boylan, W.J. (1992) Lactation Curves for Several US Sheep Breeds. Animal Production, 54, 229-233.

29. Devendra, C. (1980) Milk Production in Goats Compared to Buffalo and Cattle in Humid Tropics. Journal of Dairy Science, 63, 1755-1767.

30. Gipson, T.A. and Grossman, M. (1989) Diphasic Analysis of Lactation Curves in Dairy Goats. Journal of Dairy Science, 72, 1035-1044.

31. Dijkstra, J., France, J., Dhanoa, M.S., Maas, J.A., Hanigan, M.D., Rook, A.J. and Beever, D.E. (1997) A Model to Describe Growth Patterns of the Mammary Gland during Pregnancy and Lactation. Journal of Dairy Science, 80, 2340- 2354.

32. Ehoche, O.W. and Buvanendran, V. (1983) Yield and Composition of Milk and Preweaning Growth Rate of Red Sokoto Goats in Nigeria. World Review of Animal Production, 19, 19-24.

33. Mavrogenis, A.P., Constantinou, A. and Louca, A. (1984) Environmental and Genetic Causes of Variation in Production Traits of Damascus Goats. 1. Pre-Weaning and Post-Weaning Growth. Animal Production, 38, 91-97.

34. Mourad, M. (1993) Reproductive Performance of Alpine and Zaraibi Goats and Growth of Their First Cross in Egypt. Small Ruminant Research, 12, 379-384.

35. Madrid-Bury, N., Gonzalez-Stagnara, C. and Fiqueron, L. (1982) Milk Production of Native Goats in a Tropical Zone. Proceedings of the 3rd International Conference on Goat Production and Diseases, Tucson, 10-15 January 1982, 84.

36. Guo, Z. and Swalve, H. (1995) Modeling of the Lactation Curve as a Sub-Model in the Evaluation of Test-Day Records. Proceedings of the Interbull Annual Meeting, Prague, 7-8 September 1995, Bulletin no. 11. Department of Animal Breeding and Genetics, SLU, Uppsala.

37. Moioli, B., Astruc, J.M. and Sanna, S. (2000) Successful Establishment of Small Ruminant Recording Systems in the Mediterranean Countries. In: Development of Successful Animal Recording Systems for Transition and Developing Countries. Proceedings of the FAO/ICAR Seminar, Interlaken, 27 May 2002, 155.

CHAPTER 4

Metabolic Disorders in the Transition Period Indicate that the Dairy Cows' Ability to Adapt is Overstressed

Albert Sundrum

Department of Animal Nutrition and Animal Health, University of Kassel, Nordbahnhofstraße 1a, D-37213 Witzenhausen, Germany

SIMPLE SUMMARY

Metabolic disorders are a key problem in the transition period of dairy cows and often appear before the onset of further health problems. Problems derive from difficulties animals have to adapt to large variations and disturbances occurring both outside and inside the organism. A lack of success in solving these issues may be due to predominant approaches in farm management and agricultural science, dealing with such disorders as merely negative side effects. Instead, a successful adaptation of animals to their living conditions should be seen as an important end in itself. Both farm management and agricultural sciences should support animals in their ability to cope with nutritional and metabolic challenges by employing a functional and result-driven approach.

ABSTRACT

Metabolic disorders are a key problem in the transition period of dairy cows and often appear before the onset of further health problems. They mainly derive from difficulties the animals have in adapting to changes and disturbances occurring both outside and inside the organisms and due to varying gaps between nutrient supply and demand. Adaptation is a functional and target-oriented process involving the whole organism and thus cannot be narrowed down to single factors. Most problems which challenge the organisms can be solved in a number of different ways. To understand the mechanisms of adaptation, the interconnectedness of variables and the nutrient flow within a metabolic network need to be considered. Metabolic disorders indicate an overstressed ability to balance input, partitioning and output variables. Dairy cows will more easily succeed in adapting and in avoiding dysfunctional processes in the transition period when the gap between nutrient and energy demands and their supply is restricted. Dairy farms vary widely in relation to the living conditions of the animals. The complexity of nutritional and metabolic processes and their large variations on various scales contradict any attempts to predict the outcome of animals' adaptation in a farm specific situation. Any attempts to reduce the prevalence of metabolic disorders and associated production diseases should rely on continuous and comprehensive monitoring with appropriate indicators on the farm level. Furthermore, low levels of disorders and diseases should be seen as a further significant goal which carries weight in addition to productivity goals. In the long run, low disease levels can only be expected when farmers realize that they can gain a competitive advantage over competitors with higher levels of disease.

KEYWORDS

animal science; allostasis; autopoiesis; dairy cows; downward causation; farm management; living conditions; metabolic load; production diseases

1. INTRODUCTION

Dairy cows with high genetic merits live under quite heterogeneous nutritional and environmental conditions. Nevertheless, they are capable of adapting and maintaining their internal organization in the face of changes and disturbances, while simultaneously producing high milk yields. However, individual animals differ tremendously in their adaptive success. Adaptation to changes and disturbances in the environment requires regulations on different scales within the organism and follows a common purpose: to survive as long as possible [1,2]. Thus, adaptation is a functional and target-oriented process which can be assessed in the long run by measuring longevity. In the short run, signs of metabolic and fertility disorders as well as subclinical and clinical diseases indicate varying degrees of adaptability. When disorders occur, it is an illustration that animals have difficulties in coping with external and internal conditions, endangering their own capacity to survive [3]. As farm animals cannot escape their living conditions, their ability to cope depends on the extent their environment has an impact on them and on their level of responsiveness in order to maintain or re-establish a homeostatic state. Any stimulus (external or internal) that challenges homeostasis can be viewed as a stressor [4]. Changes in biological function occur as the animal attempts to respond to stressors and associated challenges.

Over the past decades, the focus of the dairy industry has been on maximizing milk yield, thereby creating a "nutrient highway" for nutrients from the daily feed and body reserves to travel directly to the udder to sustain milk production [5]. Indeed, farmers are not only interested in high yielding but also in healthy, long (re-)producing animals. From the farmers' perspective, production diseases are also a cost issue to which farmers can react differently [6]. For example, production diseases could be considered as undesired side effects of the production process, and factored in as associated failure costs. Alternatively, farmers could make use of investments and their work force to optimize living conditions and take preventive measures to reduce the occurrence of production disease and keep failure costs to a minimum. However, when striving for this alternative, farmers often have to face an associated increase in preventive costs.

The transition phase, defined as the period between three weeks before to three weeks after parturition [7], is the most challenging and critical period in relation to the dairy cow's health status during the lactation cycle. Major physiological, nutritional, metabolic, and immunological changes occur within this time frame as the production cycle of the cow shifts from a gestational nonlactating state to the onset of copious milk synthesis and secretion [8,9]. Cows must adjust metabolically to the dramatic increase in energy and nutrient requirements needed to ensure milk production in the ensuing lactation. Gaps between nutrient demand and supply in the first weeks of the transition period can coincide with substantial variations in the diets' nutrient content and in the daily dry matter intake (DMI), requiring comprehensive adaptation and regulation of the metabolism. Disturbances of one or multiple metabolic processes, related to the regulation of a certain metabolite in the body fluids, are known as metabolic disorders [10], and are a manifestation of the cow's inability to cope with metabolic demands [11]. Clinical diseases most related to nutritional management are ketosis, milk fever, retained placenta, metritis, displaced abomasum, and lameness [12]. The colloquial associations among various metabolic stresses and their relationships to other diseases, particular infectious and inflammatory diseases of early lactation are interrelated and have become a central focus of the interest in metabolic diseases of dairy cattle [13].

Little is known about the limitations of dairy cows in their ability to adapt to and deal with variation and disturbances in their living conditions and the gap between nutrient supply and nutrient demand. Questions arising include: how to assess what is decisive for adaptation in the specific farm contexts. Making trade-offs between various nutritional and physiological traits requires further in-depth reflections on the background, the regulatory processes involved in metabolic pathways and the possible consequences. The complexity of the issue makes it necessary to consider it from different perspectives. The main objective of this paper is to reflect on the nutritional and physiological process of adaptation to the living conditions and on the prerequisites for a successful adaptation which avoids metabolic disorders in the transition period.

2. HOW DOES ADAPTATION WORK?

To survive, dairy cows need to adapt to changing living conditions, challenged by varying gaps between individual needs and supply and by environmental changes or sudden and unforeseen disturbances. Stress response may be either non-adaptive, resulting in increased morbidity and mortality or adaptive, promoting overall fitness, emphasizing the ambivalent nature of stress. Not are only the stress factors manifold; the possible reactions of the animals to the same or to different stressors vary greatly. The adaptive responses of an individual animal may be specific to a given stressor, or general and independent of the type of stressor. Thus, adaptive success depends not only on the level and type of threat and on the responsiveness of the organism but primarily on the interactions between both.

Different terms are in use to describe the outcome of reactions in face of changes and disturbances. In this context, "robustness" can be seen as the property of being strong and healthy in constitution. The term means a kind of resistance to disturbance and irritation which prevents changes in a system's composition occurring. Thus, robustness implies endurance but not necessarily adaptation which is a special manner of being tolerant to challenges by actively monitoring disturbances and compensating for their tendencies. The term "resilience" derives from materials' sciences, meaning the power or ability to return to the original form, position, *etc.*, after being bent, compressed, or stretched. In contrast to "robustness", changes within the system's composition that occurred due to disturbances but were only temporary. In ecology, the term refers to "the ability of an ecosystem to return to its original state after being disturbed"

[14]. With respect to agricultural systems, "resilience thinking focuses on enabling a system to cope with unexpected change and disturbance" [15]. In the case of dairy cows, this term might be used in describing the ability of an animal to regain a state within a normal physiological range and/or to recover from disorders and diseases. The term "adaptation" is used in biology in relation to how living beings adjust to their environments. In the face of changes and disturbances, reactions of organisms lead to changes in the composition of their system which remain in force and are more or less permanent. According to Di Paolo [16], adaptivity is a system's capacity to regulate, according to the circumstances, its states and its relation to the environment with the result that, if the states are sufficiently close to the boundary of viability, (1) tendencies are distinguished and acted upon depending on whether the states will approach or recede from the boundary and, as a consequence; (2) tendencies of the first kind are moved closer to or transformed into tendencies of the second and so future states are prevented from reaching the boundary with an outward velocity. The specific rate of function for each organ, describing the specific range where it is able to operate, is specified by the regulatory system according to both environmental and internal conditions. A process becomes malfunctional when—due to the fact that its particular qualities limit its range of modulation—it is unable to do what the regulatory (sub)system "tells" it to do within the framework of a specific regime of self-maintenance. Diseases and disorders can be understood as negative bodily occurrences where a part of the organism fails to appropriately perform one of its biological functions [17]. Biological functions are interpreted as specific causal effects of a part that contributes to a complex web of mutual interactions, which, in turn, maintain the organization and, consequently, the part itself. Generally speaking, a well-adapted system is a set of interacting and interdependent entities, forming an integrated whole that together are able to respond to environmental changes or changes in the interacting parts.

Biological flexibility in the face of environmental demands is a fundamental prerequisite in regulatory physiological systems [18]. Moreover, regulation over a wide range of physiological and developmental conditions is required. Different concepts and terms have been coined in the past which refer to the necessary changes in regulatory processes. It is beyond the scope of this review to describe these concepts comprehensively. In short: In theoretical biology, an enduring tradition has placed heavy emphasis on the idea that biological systems employ what could be referred to as "self-determination" [19]. That is the capacity of a system's constitutive organization to contribute to the determination and maintenance of its own conditions of existence through the effects of its activity. Following Bernard's seminal work [20,21] during the first half of the 20th century self-determination was initially investigated as "homeostasis" [22] and mathematically expressed in terms of feedback loops by cybernetics [23,24]. For Cannon [25], homeostatic regulation reflects "a condition, which may vary but which is relatively constant". A common example to explain homeostasis is the ability of the organism to maintain a glucose concentration in the blood within a certain range despite considerable changes in supply and demand of glucose and in turnover rates ("stability through constancy"). To keep the glucose concentration in the blood within a physiological range, changes in the organism have to take place, regulated primarily by neuroendocrine and endocrine secretions. The need to maintain the glucose concentration within a certain range is crucial also in the context of metabolic disorders and will be discussed in further detail below. Homeostasis is traditionally seen as belonging to short-term physiological adaptations. Selye [26] was searching for a concept beyond homeostasis which also described long-term adaptations, which he called "heterostasis" and defined it as "the establishment of a new steady state by exogenous stimulation of adaptive mechanisms through the development and maintenance of dormant defensive tissue reactions". Waddington [27] suggested that homeostasis should be interpreted as "homeorhesis" which means stability of dynamics rather than stability of state. A crucial step in the theoretical development in understanding biological self-determination is the account put forward by Piaget [2]. His core idea was to integrate two inherent dimensions of biological systems into a single coherent picture: thermodynamic openness and organizational closure. In Piaget's view, biological systems are self-determining because they are organized and they are organized because they realize closure. Organizational closure is the basis of many subsequent accounts of biological self-determination [28]. One of the best-known theories is the concept of "autopoiesis" [29]. In the context of dairy production, Bauman and Currie [30] refer to homeorhesis regarding the regulatory mechanisms of nutrient partitioning and physiological processes in different stages of life history, e.g., a lactating mammal is in a different homeostatic state from a pregnant female. A similar concept ("rheostasis") involves the changes in regulation of state during circadian rhythms [31]. Knight *et al.* [32] used the terms "metabolic load" and "metabolic stress". Metabolic load was defined as "the burden imposed by the synthesis and secretion of milk", whereas metabolic stress was defined as "that amount of metabolic load which cannot be sustained, such that some energetic processes, including those that maintain general health, must be downregulated".

The term "allostasis" was introduced by Sterling and Eyer [33] to refer to changing regulatory systems ("stability through change"). Allostasis can be considered as the process of maximizing fitness in the face of environmental change as well as unpredictable challenges. Regulatory mechanisms must change in order to maintain or achieve a state appropriate for the time of day or year and in response to disturbances.

McEwen [34] developed the concept of allostasis further and coined the term "allostatic load" to describe potential permanent overburdening of homeostatic processes, e.g., by diseases. One can imagine allostatic load increasing due to the rising energetic cost of fueling regulatory processes. Accordingly, allostatic load is the sum of the energy required to maintain basic homeostasis and to acclimate to changing environmental conditions. Here, energy is assumed to be the sum of an individual's requirements (including protein, essential fatty acids, vitamins, minerals, and so on) in any given state. McEwen and Wingfield [35] extended this idea further. They see allostasis as the process of adjusting morphology, physiology, and behavior to cope with allostatic load, while allostatic state is the hormonal state required to regulate predictable changes in an individual's life cycle as well as unpredictable disturbances. Energetic demands on the individual can be further exacerbated by injury, disease, or social stress that could add cumulative allostatic load before all of the other energetic costs associated with the life cycle are considered.

Changes may occur as variation within a normal range. Predictable changes over time often emerge in an individual's life cycle. However, unpredictable events, including many potential stressors, requiring immediate physiological and behavioral adjustments may also need to be dealt with. Disturbances can occur slowly or abruptly, e.g., as change in diet, climatic conditions (heat stress), pathogen pressure, competition with other individuals that may reduce access to trophic resources, injuries, diseases, and other challenges. Effects may be cumulative over hours or days and weeks, and are disruptive or lead to abnormal activity. Additionally, current infection and disease, social status, *etc.*, may influence how an individual animal goes about its routines and responds to unpredictable disturbances. In the case of dairy cows, challenges to their adaptation can be differentiated as follows:

A group of animals or the dairy herd is exposed to the same feeding and housing conditions, although individual cows can vary considerably in their needs, due to differences in live weight, milk yield, age, or social rank. The bigger the individual gap between existing and optimum living conditions, the more cows are challenged to adapt.

Unforeseen and abrupt changes in the animal's environment, e.g., from chemical and physical factors generally concern the whole group of animals. However, individual animals may vary in their capability to react to these changes for various reasons.

Cows within a herd face individual challenges due to disturbances from ranking conflicts or due to handicaps caused by clinical and subclinical diseases or injuries.

Cows which simultaneously undergo changes in the environment and within the organism face the biggest challenges;*i.e.*, in the transition period. Considerable morphological, physiological, hormonal, and behavioral changes occur which serve in part to prepare the body for the period leading up to birth and for calving itself and also to get the cow's metabolism ready for impending lactation. These adaptations consist of mid-pregnancy anabolism followed by a significant shift to catabolism in late pregnancy and a dramatic catabolic shift in early lactation. At the same time, cows are offered a different diet than in their dry period and are often moved to different living conditions.

This paper focusses on the possible implications of metabolic stress caused by changes and disturbance in nutrition on a metabolic level. The current knowledge about the relationship between metabolic disorders and production diseases is outlined in brief before the above issue is further explored with reference to variation on different process levels.

3. METABOLIC DISORDERS AND PRODUCTION DISEASES

Desirable outcomes in farm management are cows that are successful in adapting metabolically to challenges in the transition period with minimal to no disease events, reduced avoidable culls, and efficient productive and reproductive performance. However, the reality often differs greatly. Negative side effects of the production process are obvious. They have been intensively discussed since the first Intl. Conference on Production Diseases in Farm Animals in the year 1968 which has occurred periodically ever since [36]. Traditionally the term "production disease" was regarded as encompassing the significant metabolic disorders of dairy cows (hypocalcaemia, hypomagnesaemia, and ketosis), but has been broadened in the course of time to include conditions such as a retained placenta, displacement of the abomasum, metritis, and laminitis [11]. With the increase of milk production, diagnostic method sensitivity as well as recognition of subclinical disease also increased. Dairy farms continue to be plagued by high prevalence of cow disorders, adversely affecting productivity, reproduction, and animal health and welfare [37,38]. Much intensive research has been conducted to address nutritional requirements, physiological adaptations, and metabolic associations with periparturient disease of cows in the transition stage. According to LeBlanc [39], dairy production is challenged by the fact that 30% to 50% of dairy cows are affected by some form of metabolic or infectious disease around the time of calving. Compiled periparturient disease incidence rates are presented in Table 1. According to Van Saun and Sniffen [40], most in the dairy industry still believe that there are tremendous opportunities to improve transition cow health and reproductive performance without compromising milk production, yet the solution is not clearly evident.

Table 1. Compiled periparturient prevalence of metabolic disorders from various published studies according to Van Saun and Sniffen [40].

Disease	Median Incidence Risk (%)	Range of Incidence Risk (%)
Hypocalcemia	6.5	0.3–22
Subclinical hypocalcemia	2.2	8–54
Retained fetal membranes	8.6	1.3–39.2
Metritis	10.1	2–37
Subclinical metritis	53	37–74
Ketosis	4.8	1.3–18.3
Subclinical ketosis	43	26–55
Lameness	7.0	1.8–30
Clinical mastitis	14.2	1.7–54.6
Subclinical mastitis	30	15–60

In contrast to the comprehensive data sets on the performance traits of dairy cows, the monitoring of metabolic disorders is often restricted to data from single scientific investigations. Unlike clinical diseases, subclinical metabolic disorders are even more difficult to detect; they require additional diagnostic tests, and obtaining a complete picture is not always possible. This refers not only to the severity of subclinical disorders but also to impacts which metabolic disorders might have on the development of infectious diseases and fertility disorders or mortality and culling rates, respectively. In the past, comparably few studies have been carried out to investigate relationships across a broad spectrum of health problems in relation to milk production. For example, in the case of subclinical ketosis, the reported prevalence for hyperketonemia in the first two months of lactation ranged widely from 8.9% to 34% in various studies [41,42]. Comparing current incidence rates of disease with those from past years is not helpful as records are not necessarily reliable, and cows who produce more milk than their herdmates are not automatically at increased risk of developing disorders [43]. According to Mulligan and Doherty [11] the hypothesis that high yielding cows automatically have higher levels of production diseases is likely to be as false as the hypothesis that lower yielding cows suffer from lower levels of production disease.

In early lactation, cows are in a stage of negative energy balance (NEB) caused by a rapid increase in the demand for nutrients to support milk production which exceed the increase in food intake. This results in lower blood glucose and mobilization of body reserves to provide additional energy. These processes are accompanied by elevated blood concentrations of non-esterified fatty acids (NEFA) and ß-hydroxybutyrate (BHBA) and decreased levels of calcium and phosphorus. According to Drackley [7], many health disorders in dairy cattle are attributed to the process of uncontrolled lipid mobilization in response to excessive NEB in early lactation. All of these metabolic changes increase the risk for ketosis, hepatic lipidosis, hypocalcemia, and infectious diseases, such as mastitis and metritis [44,45,46,47]. However, there is considerable variation in the plasma concentration of substrates during the day (diurnal variation) and over a period of days in early lactation. For example, plasma BHBA concentration of cows with clinical signs of metabolic disorders, generally used as an indicator for ketosis, differs widely as does the maximum plasma concentration measured in cows that did not develop clinical ketosis in the first six lactation weeks [48]. Changes in the concentrations of metabolites and hormones during the postpartum period differ remarkably amongst animals kept under similar and highly standardized conditions on a research farm, indicating that the ability to cope with metabolic stress varies considerably between individual cows [49].

Sordillo and Raphael [9] addressed the possible connections between fat mobilization and dysfunctional inflammation responses that may contribute to increased morbidity and mortality in the transition phase. While an efficient inflammatory response eliminates the invading pathogen, restores immune homeostasis, and returns tissues back to normal function and morphology, many aspects of the bovine immune system are compromised around the time of calving, especially on the level of inflammatory responses [50]. The authors conclude that the delicate balance between a sufficient inflammatory response needed for optimal pathogen clearance and the prompt return to immune homeostasis is often lost during the transition period. Immune suppression in the periparturient dairy cow is a commonly observed phenomenon and has been linked to poor metabolic status and negative energy balance [51,52]. In cows suffering severe NEB, inflammation immune genes are upregulated [53], whereas genes involved in the acquired immune responses are downregulated [54]. The mechanisms of impairment of immune defense in the transition period have been described in further detail with regard to the mammary gland [55] and the uterus [56,57]. Cows with severe metritis ate less than healthy cows in the two to three weeks preceding the clinical signs of metritis [58]. Lower feed intake is associated with increased circulating concentrations of non-esterified fatty acids (NEFA) which may directly [59] or indirectly [60] inhibit neutrophil function. Due to both high metabolic demands and pathogen challenges, cattle routinely experience substantial oxidative stress in early lactation too [61], which also contributes to a pro-inflammatory state that impairs immune defense [62]. The authors highlight that a major underlying factor in the development of transition cow disorders is metabolic stress, which occurs when cows fail to adapt physiologically to an increase in nutrient requirements needed for parturition and the onset of milk synthesis and secretion. The combined effects of altered nutrient metabolism, dysfunctional inflammatory responses, and oxidative stress can form destructive feedback loops that exacerbate metabolic stress and cause health disorders.

The role of inflammation-associated pathways in adaptation processes is not yet fully understood. The findings by Farney*et al.* [63] suggest that inflammation induced insulin resistance is, in some cases, an adaptive rather than pathological, phenomenon. Some degree of inflammation might even be required for successful adaptation. On the other hand, increased concentrations of analytes associated with stress and inflammation during the periparturient period are associated with lower milk yield and compromised reproductive performance later in lactation [64]. An underlying factor of dysfunctional inflammatory responses is the overproduction of reactive oxygen species, leading to oxidative stress [65]. While free radicals are essential for physiological processes, imbalanced or excessive free radical production plays a key role in the pathogenesis of diseases. Oxidative stress has been identified as a link between nutrient metabolism and inflammation during the transition period [62].

A close relationship between NEB and fertility disorders has been revealed and reviewed in many studies [5,66,67,68]. Increased locomotive problems are also associated with longer and more extreme periods of negative energy balance [69]. As a consequence, individual production diseases of the dairy cow should not be considered in isolation. For example, ketosis, fatty liver, retained placenta, hypocalcaemia, metritis, and displaced abomasum may be all interrelated [11]. Primary causes and the confounding factors which contribute to the development of metabolic disorders are manifold and vary considerably between farms. These factors are not only related to the feeding regime but are complicated by various management issues. Irrespective of average milk yields, some farms do well whilst others fail quite markedly in reducing clinical and subclinical problems. Vastly different nutrition and management programs produce equally good or equally poor success in relation to metabolic disorders [7]. However, little is known about the relationships between factors like body condition, feed intake, and postpartum health and performance, each varying so widely among individual cows.

4. CHALLENGES DUE TO VARYING LIVING CONDITIONS

Dairy cows live in environments that change over time. Changes that affect their self-determination and self-preservation are particularly relevant. In the first place, reactions of animals to changes in the environment are thus self-referential [70]. Cows with different and varying needs have to adapt to more or less identical living conditions. Total mixed rations are generally formulated according to the average performance of a herd or a group of farm animals. As such, rations are not adequately adapted to the requirements of individual cows within a feeding group or herd. Group size, grouping strategy, and group feeding behavior have a considerable impact on the competition between animals for feed and space and thus on feed intake [71]. Competition is perceived differently by the animals, depending to a high degree on the social rank they hold. Cows in uncomfortable conditions are less likely to use their potential for adaptation.

Living conditions are quite complex, encompassing various nutritional, hygienic, and housing spheres as well as social environment. Each sphere includes a multitude of influencing factors, differing considerably in their impact on the animals. Furthermore, they change in varying degrees over time. While climatic conditions can change quite quickly, the dimensions of space allowance, lying areas, and bunk space are more or less given conditions in a stable. Also feeding regimes on a farm might appear steady, but nevertheless can evolve more or less extended modifications in the nutrient and energy supply. Diets are composed of various ingredients, which differ not only between proportion of carbohydrates, proteins, fat, minerals, and trace elements but vary considerably in the solubility of the different carbohydrate and protein fractions and correspondingly in the availability for the animals within the digestive tract [72]. Charges of roughage or concentrate as part of the total ration differ widely in their composition. Furthermore, the proportion of single components can vary substantially due to imprecision in allocation or in mixing procedure. A comprehensive overview of the countless nutritional factors on the farm level that cause variation between and within dairy farms has recently been provided by Oetzel [12].

The amount of daily feed intake is a resultant of interactions between the composition of the diet itself, the environment in which a diet is offered and various intrinsic processes [73,74]. Feed and energy intake can change dramatically in response to changes in diet composition or metabolic state. Such changes are poorly predicted by traditional models of feed intake regulation [75]. Variation in feed intake does not only refer to the daily dry matter intake (DMI). The same DMI per day can be achieved by different frequencies and durations of eating time and meal sizes, respectively. Thus, it is not surprising that a large inter- and intra-individual variation of feed intake is observed in farm practice [76]. Even if a balanced diet following the defined average intake for a given feeding group is provided, a large variation in DMI within a group results.

When provided with free access to high nutrient dense feed, dry cows do not regulate and restrict their intake to meet their requirements and thus will over-consume energy. Dann [77] recorded intakes by dry cows that were up to 60% in excess of their requirements when fed ad libitum, with a large variation between cows. Lactating cows, provided with a suitable ration will readily consume in excess of 3.5% of their body weight as feed DM daily. This equates to between 22 and 25 kg DM/day, yet in many situations, intakes often fail to achieve even 20 kg DM/day [74]. Feed intake is depressed when highly fibrous, bulky food is fed or diets are deficient in essential components such as protein or feed access is restricted by time, bunk space, or competition, and when environmental temperatures and humidity increase and create heat stress [71,78,79]. Bunk space is most critical in farms where feed availability is limited or where overcrowding occurs. Fresh cows seem to be more sensitive to behavioral competition and intimidation, which results in altered feeding behaviors where feed bunk space is inadequate [79]. As the social hierarchy is more frequently altered, observational studies suggest a decline in feed intake, with observed greater disease prevalence [80]. The reasons for the large variation in DMI, even under standardized living conditions, are manifold, reliant on influencing factors both outside and inside the organism. Interactions between the different influencing factors create a virtually unlimited variety of combinations. What makes it nearly impossible to assess the extent of the impacts of feeding, social, and housing conditions is the fact that impacts do not emerge separately from one other. Living conditions have impacts and are challenging the animals as a whole. On the other hand, a reduction in feed intake is generally the first observed sign in cows developing metabolic disorders, particularly clinical ketosis [81]. Thus, the question always remains open, whether the feed intakes are primarily caused by impending disease, or whether the feeding patterns are induced and then result in increased disease risk?

5. VARIATION IN METABOLISM

Digestion by ruminants is the net result of a sequence of processes that occur in different segments of the gastrointestinal tract (GIT). These processes are characterized by degradation, passage, and absorption; they compete directly with one other. The sequence includes fermentation of dietary components by microbes in the rumen, acid hydrolysis, and degradation by the host animal's enzymes in the abomasum and small intestine, and secondary fermentation in the cecum and large intestine. The passage rate determines the time that feed is retained in various compartments of the GIT for digestive action and affects the nature of end-products absorbed [82]. First of all, the feed reaches the rumen, which is a large fermentation chamber, containing a complex microbial ecosystem that works in a dynamic, symbiotic relationship with the host to convert feed into energy and protein. This microbial ecosystem consists of bacteria, protozoa, archaea, fungi, and bacteriophages and is highly responsive to dietary changes [83,84]. The diverse synergies and antagonism of ruminal microbes allow ruminants to efficiently use a range of feeds. Individual cows seem to have a rumen ecosystem comprising a core rumen microbiome that adapts to different feed substrates [85]. Thus, the composition of the rumen microbiome reflects the diet, feed additives, health, age, and seasonal conditions [86,87].

To understand the dynamics of ruminal fermentation, it is necessary to understand the interactions between the rate of passage of feed through the rumen and the rate of digestion of feed in the rumen. The competing rates of fermentation and passage from the rumen dictate how much of a fermentable fraction is fermented in the rumen and how much passes undigested to the abomasum. Passage from the rumen involves mixing, disruption, and comminution of various components. The rumen operates as a partially stirring, continuous-flow reactor and different sub-compartments with different flow characteristics may be distinguished: liquid, escaping particles, and retained particles [88]. In contrast to the extensive amount of data available on substrate degradation rates, the availability of data on ruminal retention times is relatively scarce. Production and absorption of volatile fatty acids (VFA) constitute dynamic processes resulting in large postprandial variations in ruminal concentrations. Close relationships exist between production rates, the ruminal concentrations and the fluxes of VFA across portal drained viscera. The fractional ruminal VFA absorption rates increase with chain length and decrease with pH [89,90]. Diets higher in non-fiber carbohydrates (NFC) content promote the development of ruminal papillae for adequate absorption of volatile fatty acids produced during ruminal fermentation [91].

Effects of pH on ruminal VFA concentrations and kinetics of VFA absorption by the rumen wall differ considerably [89]. While removal of volatile fatty acids (VFA) from the rumen by absorption through the rumen wall are major processes that influence ruminal pH, VFA and lactic acid can build up in the rumen and reduce ruminal pH. In addition, the buffering capacity of rumen fluid is variable and is generally assumed to depend primarily on bicarbonate. Bicarbonate dependent absorption is not just a primary absorption pathway of VFA but bicarbonate can also be secreted at a capacity equal to that from saliva, thus removing protons from the rumen by neutralization. In addition, the inherent buffering capacity of the diet is involved in pH regulation, largely explained by the cation exchange capacity of feedstuffs. Low ruminal pH for prolonged periods each day can affect feed intake, microbial metabolism, and feed digestion, and has also been related to inflammation, diarrhea, and milk fat depression [92]. In face of the various impact factors, empirical models to predict ruminal pH have so far had limited success. In a recent study [93], large differences in reticulum rumen pH development were observed postpartum between individual animals when assessed continuously with an indwelling and wireless monitoring system. Considerable differences occurred despite the fact that all cows were kept under the same conditions and received the same diet before and after parturition. The results indicate that rumen pH-values are not exclusively influenced by the diet and it is thus hard to predict using experimental models; at least not on the level of the individual animal.

Fermentation rates of NDF decrease from relatively acidic to almost neutral pH [94]. The rate of fermentation of water-soluble carbohydrates and starch is, among others, affected by the amount of ruminal degradation of dietary protein [95]. Increases in the amount of ruminal degradation of dietary protein relative to readily available carbohydrates result in reduced amounts of carbohydrate stored as glycogen by the microbes, which means that the carbohydrate from the diet is fermented soon rather than stored. Another effect of increasing ruminal degradation of dietary protein relative to available carbohydrate is an increase in yield of microbial protein per unit of carbohydrate [96]. The competing rate of fermentation of starch has been reported to increase as the amount of starch in the diet increases [97]. The quantity of amino acids available to dairy cows is, among others, dependent on the ruminal degradation of dietary protein, the capacity of the rumen to use non-protein nitrogen sources for the synthesis of microbial protein, the degree to and efficiency with which ruminal degradation of dietary protein can be synthesized into microbial protein, the variation in value of different proteins to the animal that escape ruminal degradation, and endogenous protein losses [78]. Thus, there is considerable day to day variation in the efficiency of the conversion of metabolizable protein (MP) to net protein and estimates of microbial nitrogen outflow from the rumen [98]. To sum up, substantial day to day variations in digestion and fermentation processes in dairy cows cause considerable variations in the relative quantities and supply of essential nutritional elements from the intermediate metabolic processes. This supply variation represents a real challenge for the metabolism in the face of demands for milk production and maintenance.

6. DISCREPANCY BETWEEN NUTRIENT DEMAND AND SUPPLY

The average milk yield per cow has increased considerably over the last decades, primarily as the result of genetic selection. Additionally, increase in production has been supported by the improvement in environmental factors such as feed composition, feeding strategies, housing conditions, and farm management. Breeding programs have become quite successful because of the comparatively high accuracy of breeding value estimation, the moderate to high heritability of most production traits, and the use of large and fast databases containing production records of many animals and their genetic relationships [38]. The authors emphasize that increase in milk production is twice as heritable as feed intake, leading to the consequences that extent and duration of negative energy balance (NEB) has considerably increased over the last decades. During lactation, an increase in milk yield with a peak between the fourth and sixth week postpartum is functionally based on a marked increase of cell differentiation and tissue hypertrophy in the udder [99]. The number of vital mammary epithelial cells control the initial conditions for the amount of milk produced and the amount of glucose needed for the production and secretion of lactose [100]. Treating cows with bovine somatotropin could help to improve persistence of milk production because it prevents, among other things, an increase of the enzyme plasmin in the milk and thus the involution of udder epithelia. This emphasizes that the power drop in milk performance is an important programmed, endocrine and autocrine regulated process [101].

Mammary epithelial cells have an incredible level of organization and a remarkable ability to convert circulating nutrients into milk components. Cows with a high genetic performance capacity for milk production are characterized by the ability to perform intensive gluconeogenesis and partitioning of the glucose into the udder. According to Bauman *et al.* [102], the productivity of this biological factory is extensive and in terms of the use of nutrients and energy, the cow should be viewed as an "appendage to the mammary gland" rather than *vice versa*. To produce milk, the udder requires a corresponding amount of nutrients, particularly glucose, from the intermediate metabolism which correlates greatly with the amount of milk secreted. Hepatic glucose production is essential for the ultimate production of lactose which controls milk volume. The demand for glucose varies in connection with the intra- and inter-individual variation of daily milk yields. In contrast, fat is the most variable milk component, predominantly affected by nutrition. Being the major energy component in milk, milk fat percentage and fat yield can be reduced by 50% or more with little or no change in the milk yield or other milk components [102]. Milk fat is composed of fatty acids of varying chain length and degree of saturation. These are derived in almost equal manner from the uptake of circulating preformed fatty acids and the synthesis of new (de novo) fatty acids in the mammary gland. Gardner *et al.* [103] compared starch and fat supplemented rations fed to multiparous cows and noted a 4 kg/day improvement in milk yield for those fed enhanced fat levels. Fat supplement reduced milk protein content by 8% whilst milk fat levels were unaffected, but overall milk fat and protein output was increased by 9%. Computing their energy balances showed that the fat supplemented diet increased tissue mobilization.

In a study by van Knegsel *et al.* [104] cows were fed on an isocaloric and isonitrogenous basis either a mainly glucogenic (by-pass starch) or a mainly lipogenic diet. Results showed that the glucogenic diet stimulated energy partitioning towards body reserves in early lactation. Cows fed the glucogenic diet had lower NEB and reduced fat mobilization, which led to milk fat depression and less energy partitioned to milk. Thus, energy partitioning between milk and body tissue can be altered considerably by diets that differ in lipogenic and glucogenic nutrient content. High yielding cows fed with a diet enriched with crude protein (19.9%) produced an extra 5 kg milk/day of similar fat but lower protein contents in comparison to a diet with 16.7% crude protein, while ME intakes showed no treatment differences [105]. The author concluded that the extra production of milk and milk solids in high protein diets had been achieved by increased mobilization or reduced depletion of body condition. Due to a sudden increase of nutrient requirements for milk production postpartum at a time when dry matter intake and nutrient supply lag behind, nearly every high yielding cow faces the challenge of negative energy balance. NEB describes the portion of the total energy requirements of the dairy cow for maintenance and milk performance which is not covered by nutrient supply from feed intake [106]. This portion has to be provided by the organism in making use of depot nutrients to gain a balance between output and input levels. Increases in genetic merit for milk yield go together with increases in feed intake, but the latter does not fully compensate for the extra energy demands during early lactation, resulting in a more or less extended negative energy balance and increased mobilization of body reserves [107,108].

Calculated energy balance is typically mostly negative within the first 12 days postpartum [106]. Differences among cows in nadir and total energy deficits are large. Over the course of 122 lactations, mean values of total energy deficits during early lactation amounted to 1451 MJ NE_L with a standard deviation of 1062 MJ NE_L. The postpartum interval to nadir of the estimated energy balance averaged to 48 ± 29 days. First lactation cows showed a smaller energy deficit in early lactation than older cows. Moreover, cows differ considerably with regard to the partitioning of energy between different physiological systems. Thus, cows with similar energy intakes and expenditures via the milk may actually experience differences in the burden of NEB. This does not only refer to genetic make-up (e.g., high v. low genetic merit) or the stage of lactation, but varies greatly between individuals of the same genotype or in the same stage of lactation [109]. This makes predicting the dynamic of nutrient partitioning in a generalizable way (*i.e.*, across physiological stages and genotypes) more challenging.

Furthermore, emotional stress affects NEB via feed intake and glucose and free fatty acid metabolism, mainly via the sympathetic nervous system, and, to a lesser extent, via HPA axis [110,111]. Glucose is the main substrate covering energy expenditure during activities which contribute to acute emotional stress, whereas fatty acids are the main substrate which covers energy expenditure during prolonged demands. Inflammation postpartum, coupled with an increased cellular metabolism, accelerates immune gene expression while mitochondrial uncoupling further increases energy requirements and exacerbates negative energy status [112]. While numerous studies revealed that NEB can be responsible for various health disorders such as ketosis and infectious diseases [51], health problems (e.g., digestive or locomotive problems) can also be a trigger for NEB and may affect the NEB negatively in early lactating cows.

7. METABOLIC ADAPTATION

For a successful transition from late pregnancy into lactation, a cow needs to carefully coordinate metabolism across multiple tissue layers to provide sufficient nutrients and energy to support productive needs. Daily requirements for glucose, amino acids, fatty acids, and calcium for an early lactation cow are more than 2.7, 2.0, 4.5, and 6.8 times greater, respectively, than those needed for pregnancy [8]. These differences represent changes in nutrient requirements over a short period of only one to two weeks, highlighting the tremendous metabolic alterations necessary to adequately support lactation. This means that it is understandable that a cow with an inadequate ability to adapt metabolically to this transition can be the root of most postpartum diseases. Imbalance in energy and nutrient supply of high yielding cows in early lactation is inescapable. The intensity of the imbalance is influenced by both, the level of milk yield and the degree of endogenous energy provision. Levels required from the nutrient stores to compensate deficits can be theoretically calculated as the discrepancy between the total requirements and the resources deriving from daily feed intake. Generally, increase in genetic merit for milk yield is the most pronounced determinant of increased body condition loss and allostatic load [109,113]. In a study by Sutter and Beever [114], tissue mobilized over the first eight weeks of lactation was estimated to support the production of 300 L milk, from a total recorded production of 1820 L milk. Total milk production in week one was 33.8 L/day, of which 10.8 L/day (32%) were estimated to be derived from mobilized tissue. High-yielding dairy cows have been genetically selected to partition even more energy reserves into milk production with the effect that reliance on body reserves has dramatically increased [115]. Even in the case of higher dietary intake, the increased input will primarily result in greater milk production, while a similar energy imbalance remains, with no beneficial effects on body condition and reserves at all [116].

First and foremost, the mobilization requires well-functioning regulatory capacities that enable an efficient exploitation of resources, orchestrating a release of nutrients matching the requirements to a high degree and dealing with possible bottlenecks in metabolic pathways. The goal of regulation is not to preserve the constancy of the internal milieu; rather, it is to continually adjust the milieu to promote survival. The question is how the organism assesses and deals with the differences on a short- and long-term basis respectively; allowing the mobilization process to become more efficient. Furthermore, there is a question of whether the organism can rely on the regulatory opportunities available to it to prevent exhaustion due to overwhelming demands. The process of adaptation to NEB involves metabolic interrelationships between the energy storage site in the adipose tissue and skeletal muscle, the fuel processing site in the liver and the energy use sites in the mammary gland, placenta, fetus, and other tissues of the body. An increased competition for resources arises within the organism in the light of limited availability of resources. Health problems occur when these interrelationships and the adaptation process do not function appropriately to allow the cow to get through the period of NEB in early lactation.

Adaptation consists in the first place of physiological shifts in the use and conservation of body fuels: carbohydrates, amino acids and fats. The major mechanism involved is shifting the body's fuel supply away from glucose and toward the use of lipid energy sources. Lactation has an obligatory requirement for glucose that cannot be met by other fuels. Substrate-mediated events initiated by a decline in blood glucose concentration result in mobilization of non-esterified fatty acids (NEFA), which in turn results in several actions that stabilize or increase blood glucose. Besides NEFA, glycerol and proinflammatory cytokines, (TNF, IL-6) are also released [117]. Neuroendocrine and endocrine mediators are essential in the ability to adapt to NEB because fuel-mediated adaptation is not sensitive enough to function independently. Thus, endocrine actions primarily serve to control the rate of fuel metabolism. Additionally, the presence and number of specific receptors as well as their affinity play a decisive role in the mechanisms of metabolic regulation [118]. Initiation and persistence of a nutritional imbalance in early lactation in favor of the provision of external and internal nutrients for milk production is managed by simultaneous changes in the concentration of insulin, glucagon and growth hormone [119]. Low molar quotients of hormone concentrations ensure a maximum nutrient flux and nutrient uptake towards the udder. Growth hormone (GH) increases before calving [120]. High GH concentrations not only stimulate milk production but also provoke liver gluconeogenesis and lipolysis in adipocytes. Despite increased gluconeogenesis, blood glucose levels are low because of the drain of glucose to the udder. The mobilized NEFAs serve as an alternative energy source for other tissues to preserve glucose, which is used preferentially by the mammary gland to form lactose [121]. Hypoinsulinaemia promotes gluconeogenesis in the liver and acts as a massive lipolytic trigger. The udder benefits because it does not need insulin to facilitate glucose uptake into cells by the glucose-transport molecules [122]. Furthermore, low insulin concentrations uncouple the growth hormone (GH)-insulin like growth factor (IGF-I) axis in the liver because of down regulation of GH 1A receptors [123]. The resulting high blood NEFA and GH concentrations antagonize insulin action and create a further state of peripheral insulin resistance in liver, muscles, and adipose tissues [124]. In this way even more glucose is conserved to be available for lactose synthesis. Fat has been described as the "master regulator in the development of systemic insulin resistance" with good reasons [125].

Problems can occur in several steps in the adaptation process, particularly involving the adipose tissue and the liver. In the dairy cow, adipose tissue lipid accumulates during pregnancy, and catabolism begins prior to parturition and increases dramatically afterward. The duration and magnitudes of these adaptations depend on milk energy secretion, net energy intake, genotype, and endocrine environment [126]. Adipose tissue lipid synthesis is decreased and lipolysis is increased in early lactation. The magnitude and duration of these adaptations are increased in animals either consuming less energy or producing more milk. Adipose tissue is more sensitive to catecholamines in early and in midlactation and in animals with higher production. Adaptive responses of adipose tissue turn out to be inappropriate because of alterations in adipose sensitivity [9]. As adipose sensitivity increases, so does the NEFA response to a given stimulus such as catecholamine. When adipose tissue is highly sensitive, antilipolytic feedback responses may be blunted, releasing adipose tissue from metabolic and endocrine controls and allowing the rate of NEFA mobilization to go unchecked [124]. Different mechanisms lead to increased adipose sensitivity in obese cows, resulting in an interruption of homeostatic mechanisms regulating adipose mobilization. For example, high genetic merit in dairy cows is associated with increased adipose sensitivity [127,128] and a decline in circulating insulin levels [66]. During NEB, the GH-IGF axis uncouples due to a downregulation in liver

GHR and this is associated with a reduction in circulating IGF-I and elevated GH concentrations, providing an endocrine environment that promotes the direct action of GH on lipolysis and gluconeogenesis in early lactation [129]. The adipose levels in a cow are also a critical factor affecting adipose sensitivity. Studies conducted by Rukkwamsuk *et al.* [130,131] revealed that cows on a high-energy diet ad libitum throughout the dry period had adipose tissue that was less responsive *in vitro* to the antilipolytic effects of glucose and ketone bodies compared with restricted-fed controls. This reduces the effects of fuel-induced, autoregulatory negative feedback signals on lipolysis, allowing increased NEFA mobilization in obese animals. In response to NEB, over conditioned cows mobilize more body fat than thin cows and subsequently are prone to develop metabolic disorders [132]. Obese cows have less effective insulin feedback regulation in the adaptive response to NEB and greater adipose sensitivity and NEFA mobilization compared to leaner cows [133]. This can result in inappropriately high blood NEFA concentrations that overwhelm the capacity of the liver for fuel processing and interconversion.

Recent findings revealed the effect of dietary level of energy during the dry period on feed intake, energy balance, milk production, and composition [134]. Cows being fed a diet exceeding energy requirements during the whole dry period showed an increased incidence of hyperketonemia while a controlled-energy dry period feeding approach was successful in minimizing the degree of negative energy balance postpartum as well as hyperketonemia. Several studies have shown that the energy level and composition of the pre-partum diet also have an impact on lipid metabolism, liver function, and the overall health of the cow post-partum [135,136]. A transient decline in high density lipoproteins and loss of this molecule's anti-inflammatory and antioxidant actions may account for a portion of the increased oxidative stress recognized to alter immune function in transition cows [9,137]. Besides fat, protein also plays an important role in the provision of glucose. Absorbed and mobilized amino acids from the muscles are partitioned to either gluconeogenesis or ketogenesis (oxidized for energy), milk protein or body (and conceptus) proteins. Although utilization of propionate for gluconeogenesis is extensive, amino acids have the potential to increase in importance as gluconeogenetic substrates when glucose demand is substantially increased [138]. Van der Drift *et al.* [48] observed large variation in the onset and duration of periparturient fat and protein mobilization. Cows mobilized in total 35% ± 26% of back fat thickness and 18% ± 9% of longissimus muscle thickness when kept under similar conditions at an experimental farm. Ketone body production was restricted in cows that mobilized more muscle protein relative to fat tissue after parturition. The relative partitioning amongst these pools depends on the mix of nutrients entering the body, the availability or need to replenish body tissue stores, and other physiological influences on the cow (stage of lactation, phenotypic, and genetic capacity for milk production, environmental conditions, age, growth stage, previous nutrition).

The hepatic response to NEB is to produce glucose and ketone bodies and take up large amounts of NEFAs from the blood for mobilization as other forms of energy. Apart from the activity of hormones, metabolic regulatory events in the liver are directed primarily by the level of metabolites, either in excess (e.g., free fatty acids) or shortage of supply (e.g., glucose) [139]. Reynolds *et al.* [140] reported major changes in the hepatic flux of NEFA as cows progressed from a pregnant state into lactation. Measured NEFA flux at day 19 before calving was equivalent to approximately 1 mol palmitate equivalent/day and a similar level was recorded at day nine before calving. In contrast, by 11 days post-calving, NEFA flux increased to 5.5 mol palmitate/day and whilst the level declined as lactation progressed, it was still almost twice the pre-calving baseline at lactation day 83. Maladaptive responses in the liver can arise at several points in the metabolic processes. Comprehensive reviews on the pathophysiological basis of ruminant adaptation to NEB and the development of ketosis and fatty liver is provided by Herdt [141], Drackley *et al.* [47] and Ingvartsen [112]. In short: Type I ketosis occurs when the demand for glucose outstrips the capacity of the liver for gluconeogenesis. Pathways for gluconeogenesis might be at the height of stimulation, but the supply of glucose precursors is insufficient to permit maximal glucose production. Thus, glucose production is limited by substrate supply in the first place.

Type II ketosis occurs when large amounts of NEFAs are delivered to the liver, but gluconeogenesis and ketogenesis are not at the height of stimulation. Mitochondrial uptake of NEFA is not as active as in type I ketosis. NEFAs not used for ketone body synthesis are esterified in the cytosol, forming triglyceride. Transport of triglycerides from the liver to other tissues requires the synthesis of very low-density lipoproteins (VLDL). Secretion may be near disabled when blood NEFA concentrations are high, leading to an accumulation of triglycerides in the liver. As the degree of fatty infiltration increases, normal functions of the liver are adversely affected. For example, the development of a fatty liver can interfere with hepatic gluconeogenesis capacity [142,143]. The ensuing hypoglycemia can lead to further adipose mobilization, creating a destructive spiral of events leading to fatty liver of increasing severity [141]. Fat infiltration also impairs the ability of the liver to detoxify ammonia to urea [144]. Ammonia decreases the ability of the liver to convert propionate to glucose [138]. Fatty liver and type II ketosis are usually seen clinically shortly after parturition, well before peak milk production. This reflects the association with adipose sensitivity [145]. Fat accumulation in the liver begins to occur in the last weeks of gestation, as adipose sensitivity increases. Factors influencing adipose sensitivity and energy balance in the late dry period are important concerns in the prevention of fatty liver and type II ketosis.

8. HIERARCHICALLY ORGANIZED AND NESTED SYSTEMS

The previous explanations followed the processes during the flux of nutrients from the environment through the digestive tract, and processes within the intermediate metabolism on different scales. As stated above, transformation processes are influenced by countless variables on each scale. Changes in perspectives from a top down to a bottom up approach and *vice versa* might help to improve our understanding of what restricts cows' metabolic capabilities to adapt to nutritional changes and to grasp metabolism with a genuinely systemic approach. Ideally metabolism should be expressed mathematically, but this is made difficult by the fact that its elements are simultaneously both variables and functions [146]. Metabolism is the inherent process of living organisms to obtain nutrients and energy from the feed ingested for self-referential purposes. Feed is made up of substrates (proteins, carbohydrates, and fats, *etc.*), degraded by enzymes in the digestive system into absorbent nutrients and fuel. The body can use the nutrients and the fuel right away, or it can store them in body tissues, such as body fat and muscles. A metabolic disorder occurs when abnormal chemical reactions in the body disrupt or impair the process of utilization. When this happens, the animal might have too much of some substrates, too little of other ones, or too much and too little simultaneously so that proper function is disturbed. A metabolic disorder can develop also when some organs become diseased or do not function properly. Thus metabolic disorders can be divided to structural and functional components. A structural component is represented by the arrangement and configuration of compartments (organs, tissues, cells). Details of a structure can vary considerably between animals, e.g., in relation to the proportional size of sub-systems and their functionality. For example, the large variation in number and form of villi in the rumen, and the condition of other semipermeable borders which have a considerable effect on the flow of nutrients between the compartments [147], illustrate structural variability. Functional components contribute to self-regulatory mechanisms, responsible for the stability of the body's internal environment [148]. Their effects can be regarded as dynamic constants, including blood metabolites, osmotic pressure, or pH amongst others. However, metabolic and external influences cause permanent or periodic changes in parameters of the internal environment. Therefore, the mode of activity and parameters of results of metabolic functional systems change consecutively. Self-regulation of functional systems is the mechanism which maintains these parameters at optimal levels.

Changes in the amounts and patterns of nutrients flowing through body compartments where these nutrients are degraded, synthesized, or stored, reflect the large day to day variation in nutrient availability the metabolism has to deal with. Metabolic disorders primarily develop when discrepancies exist between what is needed to sustain a certain balance and what is supplied from various sources, and when the organism cannot cope appropriately with these discrepancies. Discrepancies are not limited to single substrates but encompass a wide range of factors and processes, including interactions between different substrates within and between different compartments on different scales. The complex systems in living organisms operate both on a level of time and of space. The hierarchy of systems or scales is introduced because it corresponds to our subjective views of a system based usually on our various discrete experimental viewpoints. In their famous work, Oppenheim and Putnam [149] conceived of nature as being constituted by a hierarchy of objects that in turn, defined a hierarchy of distinct sciences. In agricultural and animal science, multiscale approaches inspired from theoretical physics have been developed in an essentially unidirectional (bottom-up) approach to integrate parameters at a given scale into reduced reproduction at higher scales [150]. However, lower-scale properties are also directly coupled with properties at a higher scale. The very complexity of living systems and biological functions lies partly in the presence of these bi-directional feedbacks between the upper and lower scale. An illustration of the hierarchical structure of living (autopoietic) systems is presented in Figure 1.

Based on basic biological considerations by Maturana and Varela [1] concerning the characteristics of self-organizing autopoietic systems, reworked, and reformulated by Razeto-Barry [70], the vertical structure of living systems in agriculture can be also described as a hierarchy of scales. An example is provided in Figure 2 with the basic level of cells (1); interacting with material elements, and organizing themselves in tissues and organs (2) as superordinate systems, integrated in a whole organism (3); which is part of a herd (4); while the herd is an intermediary level between the single animal and the agro-eco-system (5). Microbiologists would probably see the genome as the basic level [150], while agriculturalists might extend the hierarchy of systems to far larger scales [151]. In the current context, it is primarily important to bear in mind that each superordinate system acts as the environment for the corresponding sub-system. It is the duality of the perspectives in particular which enables a more comprehensive understanding of biological processes. Without the whole living organism, its parts—the organs and the cells—could not exist or be sustained. Thus, the living system continuously produces the components, which constitute it by itself. Furthermore, these components steadily sustain and regenerate the superordinate system [152].

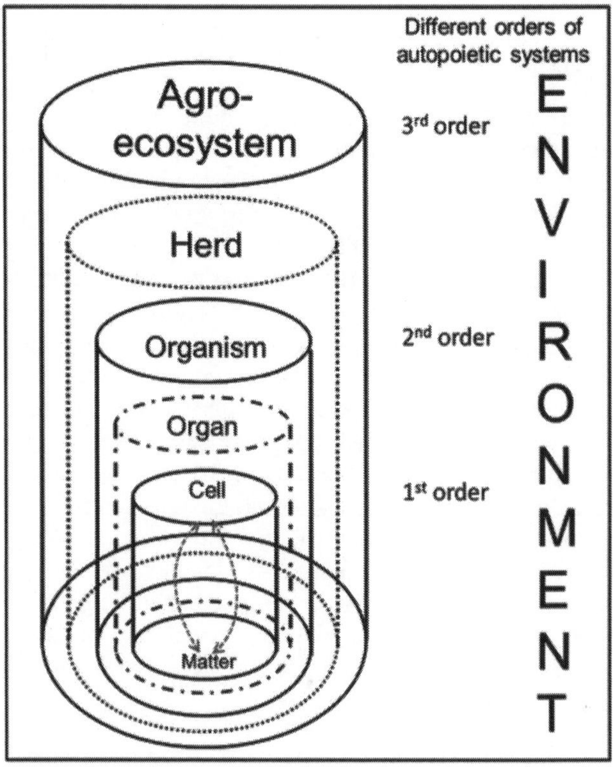

Figure 1. Hierarchical structure of autopoietic living systems of different orders within an agro-ecosystem.

From a biological-systemic perspective, an organism is an integral and functional system in continual process of exchange with the environment. It represents an entity with a functional integrity [153]. Maintenance of the entity is an active process of self-organization which requires resources. In striving for maintenance, external material is regulated by the living systems following to their own principles and purposes, while integrating and transforming the material within the system. This principle is valid for the single cell as well as for the organism and for the farm as an agro-ecosystem. For instance, cells adapt their own metabolism and their functional capacity to the availability of external resources surrounding the cells [154]. To a certain degree, cells are able to perceive the availability of nutrients directly and transform this information into modified metabolic flux rates. Thus, processes are not exclusively determined by internal dispositions (e.g., genome or proteome) and conditions (e.g., anabolic or katabolic metabolism) or by impacts of the physical environment (e.g., availability of nutrients or external stressors). Living systems react due to the way the system is configured and adapt constantly to their environment. They select the appropriate components from the environment and give them a meaning of incentives, objects of perception, and interests [152].

Nutrients that would enable the individual animals to meet their requirements efficiently and comprehensively on a daily basis are neither available in the wild nor in livestock farming. Animals can rely on different metabolic pathways by which nutrient deficits or surplus can be compensated for an extended time period. Ability and skills to deal with nutrient imbalances are decisive if the animal is to succeed trying to adapt to inappropriate and varying supply conditions. Metabolic and reproductive disorders, diseases, and mortality indicate that the cows' ability to adapt to variations in its environment is overstressed due to disturbances in the interdependency between super- and sub-ordinated systems and their dynamic transition processes. Disorders and diseases can be understood as negative bodily occurrences where a part of the organism fails to perform one of its biological functions appropriately. This interpretation corresponds with the concept of biological malfunction, recently described by Saborido and Moreno [17]. In this concept, biological functions are interpreted as specific causal effects of a part or a trait that contributes to a complex web of mutual interactions, which, in turn, maintain the organization and, consequently, the part itself.

Change and disturbance occur both between and on all scales. Due to a large variation in composition and ingredients of: feedstuffs, diets, daily feed intake, as well as digestive and absorbent capacities, a variable supply from the intermediate metabolism meets a variable need in nutrient and energy requirements in the various compartments of the body. The gap between demand and supply has to be covered by regulatory processes which for their part consist of influencing factors with respect to input, partitioning, and output variables. Dairy cows differ considerably in their adaptability to nutritional changes and metabolic disturbances, particularly to NEB. Last but not least, this is evident through large differences in prevalence and incidence of metabolic disorders despite similar environmental conditions. The reasons behind the differences in adaptability are manifold. How to evaluate the degree to which adaptation occurs in this specific situation, taking into consideration the large intra- and inter-individual variation in the nutrient flow that takes place on and between various scales, is therefore of particular interest. This is also true for various other factors involved in the adaptation process. The large individual differences in input, partitioning, and output variables indicate a variability of metabolic processes on various scales that cannot be narrowed to a single factor but which needs to encompass the interconnectedness of variables and the nutrient flow. Added to this is the fact that most problems which challenge organisms have a number of equivalent solutions, which can be thought of as existing in a vast neutral space [155]. A neutral space is a collection of equivalent solutions to the same biological problem, embodied in the biological systems ensuring an organism's survival. This is the case for the multiple levels of biological organization. Understanding the structure of neutral spaces is critical to understanding the robustness of biological systems and to what extent the robustness itself can evolve.

A limited availability of nutrients and energy provokes severe competition between different tissues in the need for nutrients to sustain their various functions. Limitations require partitioning, and partitioning requires prioritization in guiding the nutrient flows, and ensuring that the demands of other cells, tissues, and organs within the organism are not completely neglected [30]. Thus, there is a need to prevent ruinous competition between sub-systems, to avoid being swamped by unwanted side reactions which clog the system or parasitic reactions by single organs to the expense of other organs which may cause the whole organism to collapse. Nutrient prioritization in early lactation to favor milk production over fertility is a reasonable strategy in biology [156]. As nutrition becomes scare, the lactating dam will preferentially invest the limited resources in the survival of living offspring rather than gambling on the oocyte that is yet to be ovulated, fertilized, and cared for during an entire gestation. Selection of high milk yields takes advantage of the genetically programmed readiness of the dairy cow to enter into a negative energy balance at the onset of lactation and to mobilize resources from its body tissues. However, selection has advanced into dimensions that are far beyond the initial intention to ensure nutrient supply to the off-spring via milk. Thus, there is no basic conflict between milk production to safeguard the off-spring on the one hand and self-preservation of the dam on the other. Life-threatening conflicts occur, when the gap between demand and supply gets to the stage where metabolic regulations are at risk of failing to mobilize the bodily resources needed to compensate for the deficits between nutrient output and intake. In case of failed regulation, the maternal ability to nourish the off-spring can be completely lost due to the death of the dam.

Adaptation depends *inter alia* on the degree to which the individual requirements of cows are met through the structure and organization of the farm system as the individual peculiarities of dairy cows mean that they differ considerably both in their general requirements and respecting their specific performance within the lactation course. To fulfil these demands, the farm system itself is in need of appropriate resources (high quality feed, investments, labor capacity, knowledge, *etc.*). What is needed on individual farms cannot be generalized but also depends on the specific farm system. At the same time, farmers faced with low milk prices are striving and fighting for the sustainability of their own farm system in a global market where all dairy farmers produce more or less the same commodity, thereby placing high pressure on production costs.

9. CHALLENGED FARM MANAGEMENT

Intense selection for milk production has resulted in an immense priority for the high-producing dairy cow to partition energy to milk, at the cost of body reserves. This has resulted in an excessive negative energy balance (NEB) and high prevalence of metabolic disorders and poor reproductive performance. Thus, milk production, animal health, and reproductive performance in high-producing dairy cows generate conflicting interests. Traits in relation to cow health have only recently been included in the selection criteria on the basis that considerable genetic diversity with respect to health and reproductive performance exists in the current population [67]. However, functional traits have a comparably low heritability thus questioning whether such approaches will bring the desired improvements in the foreseeable future. In the meantime, animals will continue to suffer from production diseases while milk yields will further increase and linked with the negative side effects. The preparedness of the dairy cow to enter into a negative energy balance in early lactation while mobilizing nutrients from body tissues is an essential precondition for a high milk performance in the total lactation period. Apparently, dairy cows do not possess an effective feedback loop, e.g., to reduce milk yield and depletion of body reserves to a degree that prevents metabolic disorders and does not compromise the survival of the cow itself. Thus, it is the responsibility of the farm management with their various organizational skills and resources to avoid overstretching the gap between demand and supply. The crucial question is: when is the border which endangers the ability of individual animals to sustain themselves crossed? As this paper's concept explains, such thresholds can neither be generalized nor predicted for the individual animal.

As milk yield increases, the management challenge of meeting a cow's dietary nutrient needs becomes greater. This implies that increases in milk yield achieved through genetic selection requires improving the entire availability of nutritional resources. However, not all high yielding cows are capable of achieving an adequate energy intake, especially under grass-based management systems [157]. While energy requirements have increased more rapidly than dry matter intake (DMI), the nutrient density of diets has increased simultaneously, enhancing the risk of subacute ruminal acidosis [158]. Although advancements have been made in feeding practice to minimize the risk of metabolic diseases, the periparturient period continues to present some of the greatest challenges in the animals' health [107]. Nutritional approaches vary enormously between farms and in many instances are failing to meet the more exacting demands on modern dairy cows [159]. Too many farmers are still concerned about feed costs per ton when the real focus should be on feed cost per liter milk produced. Larger herds usually result in more cows managed per available staff member. Declining availability of qualified dairy staff has added concerns that cows are receiving less individual attention, which is likely to have contributed to the high level of metabolic and fertility disorders. Dry cows are still often considered as second-class citizens or nonworking cows and thus receive little attention and poor-quality forage. Separation of dry cows into an independent group or into sub-groups is not readily practiced. Dry cows are often housed in a low-producing group [67].

Given that nearly all cows experience reduced feed intake and loss in body condition postpartum, there is a large variation between cows in the extent and duration of negative energy balance. Because the energy balance calculation requires knowledge of feed intake and digestibility, milk composition, and weight; in farm practice, energy balance values can often not be calculated due to lack of data on these components. Correspondingly, farm management is often eager to follow general recommendations instead of trying to adjust feeding regimes to the degree of NEB and the variation between cows, as this is easier. In the light of highly variable initial and boundary conditions in farm practice and limited availability of resources in dairy management, it is more than questionable whether "best practice" is still a realistic and valid concept. There is much evidence that specific measures operating under certain conditions are not necessarily effective, when compared with others in achieving the same purpose. Feeding higher-concentrate diets in close-up groups to meet increasing energy needs, is often practiced but recognized as decreasing the feed intake due to high availability of propionic acids which provokes satiety [73]. This situation can be counterproductive at a time when cows are accelerating their milk output, which accelerates lipid mobilization and results in associated negative consequences. On the other hand, excessive intakes of energy relative to requirements in late lactation and dry period is almost certainly expected to increase the risk of "fat cow" or "fatty liver" syndrome [77] and increase the risk of insulin resistance in cows and to elicit changes similar to obesity and type II diabetes [160]. Besides the availability of energy, the relevance of the dietary provision of metabolizable protein has been emphasized, based on strong indirect evidence backing the importance of the maintenance of maternal protein stores on long-term health, productivity, and reproduction [119].

With the continued challenges to cows navigating the transition period, some researchers have considered altering the dry period to minimize metabolic changes. Shortening or even skipping the dry period improves dry matter intake peripartum, reduces milk production in early lactation, improves energy balance, and reduces the number of days postpartum till resumption of ovarian activity [161,162]. A recent meta-analysis of 24 studies manipulating dry period length showed lower milk production, improved energy balance, and decreased risk of ketosis, but no difference in other diseases or fertility [163]. In a recent study by Chen *et al.* [164], cows with a 0-d dry period had an improved metabolic status in early lactation, indicated by lower plasma concentrations of non-esterified fatty acids, greater plasma concentrations of glucose, insulin-like growth factor-I, insulin and lower mRNA expression of pyruvate carboxylase in the liver, compared with cows with a 30-d or 60-d dry period. However, adoption of shorter dry periods has been limited because of the milk loss concerns.

A way to reduce the imbalance between nutrient supply and demand in early lactation is not only to improve the first but also to temporarily decrease the latter. While most dairy cows are milked twice daily, it is not uncommon in intensive dairying systems to increase milking frequency up to four times daily to increase milk production. Reducing milking frequency is much less common. However, the overall energy balance of cows during early lactation is improved with one-daily milking [165]. In dairying systems where an emphasis is placed on animal health rather than on milk production per cow, this practice might fit well under certain infrastructural farm conditions. Depending on the stage of lactation, breed, and parity the reduction in milk yield losses varies considerably and can amount to approximately 22% [166]. This figure has been revealed as an average milk loss across 30 different international short-term studies. Production expressed as milk solids (*i.e.*, combined fat and protein yields) is also lower with once-daily milking, despite the fact that milk fat and protein content of the milk increase [52,167]. The increase in fat and protein content is not sufficient to compensate for the decrease in milk volume. However, it may provide a tool to manage the metabolism and energy balance of cows during early lactation better [168], e.g., at highest negative energy balance period. Results from a study by Carbonneau *et al.* [169] revealed that blood concentrations of glucose were higher and concentrations of non-esterified fatty acids and ß-hydroxybutyrate were lower in partially milked cows compared to control cows during the treatment period. In other studies too, cows milked once a day showed an improvement in their metabolic profile [52] and immune function [59]. In contrast, cows subjected to an increased milking frequency were 1.4 times more likely to be classified as subclinically ketotic as the control cows [170].

One-sided selection for high milk yield is likely to produce cows that are increasingly more vulnerable to disease and poor fertility [171]. The authors concluded from different studies that negative genetic correlations between production and fitness traits increase in less favorable environments and that selection for increased yield has decreased the adaptability of modern cows. Keeping such animals in living conditions where farmers have increasingly less time available per cow demands even higher management skills. In situations of inferior management, unfavorable genetic correlations between milk yield and metabolic disorders are expected to be stronger than in herds with superior management. In a study allowing discrimination between the roles of genotype (G), environment (E) (e.g., feed caloric density and milking frequency), and GxE interactions, the effects of genetic merit and milking frequency were significant only in the groups that were fed rations with high caloric density [109]. However, signs of severe negative energy balance, poor protein balance, and low body condition scores were not concentrated in the highest producing cows. A lower feed caloric density and extra milking had strong unfavorable effects on both energy and protein balance, and emphasize the possible effect of mismanagement on animal health risks. Increased milking frequency increased milk yield, although only significantly when the cows received high caloric density rations. Milking the cows three times a day instead of twice significantly decreased energy balance. Thus, environmental factors like feeding management and milking frequency strongly influence the cows' energy balances. High genetic merit cows seem more prone to allocate resources toward high fat output in times of nutrient-deficient diets and toward high protein output with high nutrient diets. High genetic merit for milk yield seems intrinsically connected with the allocation of resources from maintenance toward milk. To prevent compromise of animal health status; demands on the provision of appropriate feed and feeding conditions; farmers' time and management skills are clearly increasing. In a direct comparison of low- and high-yielding genetic lines of cattle, at relatively similar levels of dry matter intake, there is no evidence of a difference in the fractional recovery of DE from intake energy or ME from DE [172]. Similarly, no differences were found in the efficiency of ME use for milk energy. Thus, increase in productivity due to higher milk performance per cow is based primarily on the proportional reduction in nutrient resources needed to cover the demands for maintenance. In terms of maintenance requirement, the high yielding cow is more efficient as maintenance cost per unit milk is diluted [173].

Against an industry-held view that milk output is the single most important aspect in an efficient dairy business, it is contended that producers need to refocus on factors which also affect profitability, mainly: nutrient efficiency, cow health, fertility, and longevity [6,174,175]. In animal nutrition, the most efficient use of nutrient resources is generally obtained when each animal is supplied with energy and nutrients according to their individual demands [40]. Both exceeding supply before calving and under-nutrition postpartum not only reduces efficiency but requires additional resources of regulation to maintain the internal processes within a certain range of homeostatic conditions [9,176]. Feeding farm animals according to their current requirements regarding nutrients and energy provides the best orientation for the farmer to gain a high efficiency in the use of nutrients while simultaneously reducing the risks of metabolic disturbances. However, to meet such a goal requires additional resources, amongst others: high quality feed, good size and type of stalls, suitable amount of eating space available to the cows or ventilation systems in place, and last but not least knowledge and skills to reduce possible gaps and compensate for deficits [12]. Thus market conditions which do not provide enough income for farmers and do not fully cover the expenditure during the production process threaten not only the sustainability of the farm systems but also that of the dairy cows and their functional integrity of organs and tissues through downward causation.

While farmers have to consider changes in the availability of resources, they can no longer ignore consumer concerns about animal health and welfare, and food safety. Increasing disadvantages of the intensification process and conflicts with the interests of other stakeholders call for a critical assessment of the sustainability of dairy production [177]. What is sustainable for agricultural systems is subject not only to system-inherent and self-referential demands but also to changes in the market. Thus, farm managers are not only agents of control but components of the agricultural system itself. The fact that market prices and consumer preferences are primarily dominated by forces beyond the farmers' control makes the issue even more complex and somewhat confusing. In providing the essentials for remaining both competitive and sustainable, the farm management is restricted to only a few options; either to further reduce production costs per product unit or to obtain a higher selling price. As long as qualitative traits such as the production of "healthy" food from healthy cows are not honored by premium prices, farm management has to focus on how to improve the overall efficiency of production processes [178]. This cannot be achieved solely by addressing the milk yield of dairy cows. In fact, farmers have to take into account the cost-benefit relationships on various scales, including failure costs caused by metabolic disorders and production diseases as well as the preventive costs needed to reduce these. Expenditures for preventive measures can be seen as additional input to reduce monetary losses of disorders and diseases [6]. The higher the preventive costs, the lower the failure costs and *vice versa* (see Figure 2). If no preventive and control measures are taken, the losses due to disease are at their maximum (Lmax). With maximum prevention, the failure costs due to disease will be at a minimal

(Lmin). Because the relationships between prevention and failure is not linear, there is an optimal level of control (Lopt). To gain profit from reduced prevalence of production diseases, the farm management has to strive for an optimum between failure and preventive costs. Due to the various and highly varying factors involved in the development of productivity and in the development of disorders, and due to the large differences in the availability of resources mentioned above, such a balance is different for each farm system. Furthermore, the balance varies in relation to changes in the environment of the farms, especially regarding changes in food and feed markets.

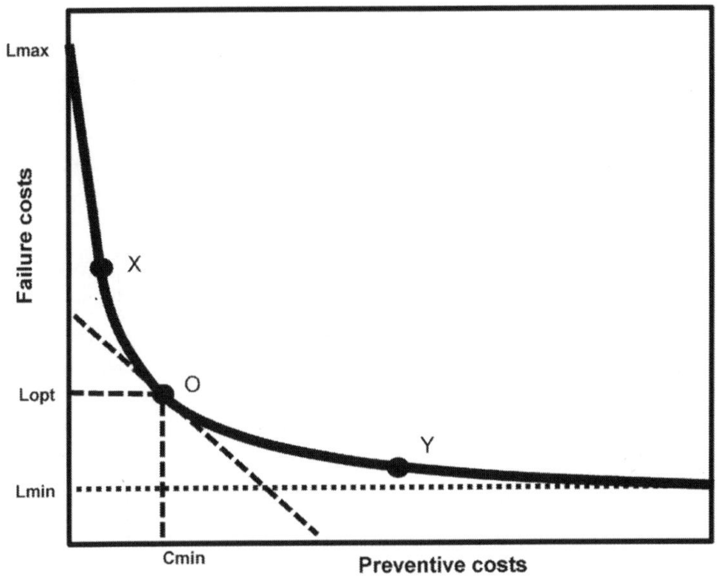

Figure 2. Schematic representation of the relation between failure and preventive costs [6].

It is in the nature of much-circulated general recommendations that they only marginally address the context in which the biological processes take place. Correspondingly, they may appear counterproductive, especially when they distract the farm management from looking for farm specific solutions. The latter should rely on valid data which unfortunately are often not available to the desired quality, probably because data acquisition itself is very time consuming and reflecting upon data needs much mental energy [179]. Moreover, changes in perspectives are required to implement and assess bilateral adaptation processes, on the one hand by the animals towards their living conditions and on the other by the farm management in relation to the specific demands and limitations in the adaptation capacities of dairy cows, to the benefit of both animals and farmers.

10. CHALLENGED AGRICULTURAL SCIENCE

Agricultural research and animal science as a sub-discipline deal with agricultural systems and with socio-technical and socio-economic circumstances as well as the superordinate systems which influence them. Criteria for assessing the quality of agricultural research are shaped on a problem-oriented systems research basis [180]. This consists of the close connection between basic and applied research, the multi- and interdisciplinary character of agricultural research and the orientation of research towards the solving of issues relevant to society. Regarding the high prevalences of metabolic disorders and associated production diseases in dairy production, agricultural science so far cannot claim to have been able to reduce health problems. Despite tremendous scientific efforts in the past, the situation has not improved adequately. Animal scientists might at least claim that without technical and scientific progress in gaining more insights and knowledge about the pathophysiological processes, the situation might have become even worse. However, lack of success in reducing production diseases does not meet goals set by agricultural scientists and there is thus an impetus to reflect on whether the previous approaches in dealing with health problems are still appropriate to the questions on hand. In the first instance, a reductionist approach has been employed, accumulating large quantities of detailed information with little attempt to incorporate it into a broad view of the whole subject. Thus, scientific efforts have provided a wealth of knowledge but to a large extend, this information has not been integrated into coherent biological models. Members of animal science in particular have not done well to integrate output productivity and negative side effects with respect to animal welfare [181].

In the case of adaptation to metabolic challenges too, there is a predominant tendency to reduce the differences between animals' ability to cope with nutritional and metabolic challenges to single factors, with a priority to differences in the genome (e.g., [47,182,183]). Based on the methodological advances made in the human genome project, and the need to interpret findings in a more biologically meaningful fashion, "system biology" has emerged as a new approach [184]. In the meantime this has been applied to many species, including dairy cows. The pace at which transcriptomics has expanded has forced the discipline of genomics to embrace the notion of holism as far as animal function is concerned [185]. The new approach focusses on the gene/protein expression (proteomics) on the cell/tissue level and the associated changes in biological functions. It tries to combine omics-generated data with mathematics and computer science to gain knowledge of biological behavior of a cell in either a bottom-up or top-down direction [186]. Recent developments in technology have paved the way for new modelling approaches at various levels of complexity [115,187,188]. One of the main objectives is to be able to predict the ability of an animal to respond to the nutritional limitations that arise from the environment in which it is placed. The perspective is that of incorporating increasingly detailed descriptions of genotype via genomics, and of the downstream metabolic machinery via gene expression, proteomic and metabolomic information [115]. As indicated by the authors, this remains a daunting task; not only due to serious methodological hurdles. The authors also acknowledge the danger that even if all the omic information were available, it might be still too complex for the minds of scientists. Whilst generating biological information about dairy cattle through the application of genome-enabled tools is no longer a bottleneck, the biggest challenge is to interpret findings in a more biologically meaningful fashion and eventually to communicate biological knowledge systematically by linking the genome to the whole organism [188]. The authors admit that the use of the omics' tools alone cannot provide a full picture of the biological system and can thus be misleading when interpreting biological outcomes, while scientists are still struggling to interpret the results from any one of the omics' technologies in a holistic and unbiased way.

While many genome-wide association studies have frequently shown that they do not explain the majority of the variation in whole-animal phenotypes, it has become clear that the relationship between the genome and the phenome is best characterized in terms of causal interdependency [150]. Genes are often found to possess multiple functions, which are sometimes critically dependent on the context and the specificity of enhancers for promotor sequences is often surprisingly loose [189]. Thus, even for fixed genotypes and environments, a large variability of phenotypes can occur across a population [190]. According to McNamara [181], "systems biology" means different things to different people. At minimum, it is the recognition that each piece of the system has a specific function related to the outcome of the entire system, not just the subsystem in which the molecule acts. For Cornish-Bowden *et al.* [146], the term "system biology" is little more than a euphemism for gathering even more details on an even larger scale, and not, as it should be, the study of biological systems as systems rather than as a collection of components. The more complex the level at which one seeks to explain a living system, the greater the need to examine the network of interactions.

In opposition to the omics' approach, it might be further remarked that in search for general principles and generalizable regularities, the mathematical approach is faced with the tremendous variation that occurs on the nutrient pathways from nutrient intake to utilization and secretion via milk. The variation is increased further by the interconnectedness of the variables on different scales, organized by various forward and feedback loops. As origin, scales of time and space, control, and functional significance of fluctuations in biological systems are largely unknown, it is questionable whether the genomic approach will be able to achieve biologically meaningful aggregated phenotypic descriptions in a foreseeable time period, let alone provide a valid base for decision-making processes on a farm level.

It has become increasingly clear that understanding nutrient partitioning is central in grasping the complexity of adaptation to shortages and imbalances in nutrient supply in relation to requirements, and thus of the ability of dairy cows to cope, especially when challenged with abrupt and unforeseeable changes. This refers not only to the nutrient partitioning within the animals but also to the level of the farm. In looking for approaches to gain knowledge in predicting nutrient partitioning in dairy cows, animal scientists often recognize the central role of the genotype. This however, leads to the question, how a valid prediction regarding the reactions of animals with a high variability in their medical history towards more or less unpredictable challenges can be obtained. Although the degree of deficits might be comparable between cows, for instance in the extent of NEB, there are large differences in the possible reasons behind the deficits and thus in the reactions and adaptation processes of the individual animals as outlined above. In the light of metabolic disorders and production diseases, affecting more than every other cow in early lactation to a varying degree, expectations concerning success in gaining predictions on the capacities to adapt are—to formulate it cautiously—astonishing, at least from the perspective of animal nutrition and animal health. Probabilities for successful adaptation can be at best predicted on the population level. This however, does not necessarily mean that what might be expected on the population level is valid for the corresponding farm level where adaption takes place in reality.

To look beyond narrowly molecular perspectives for answers to biological questions, it is important for metabolic adaptation not just to look at the total sum of the individual parts, but how they interact, and especially if they interact successfully in preventing the animals from suffering metabolic disorders and associated production diseases. A widespread view is that causation operates in an exclusively bottom-up way, from the microscopic to the macroscopic. Adherents of such a view often express perplexity at the idea that causation might even run in the reverse direction [150]. In tracking the flows of causal influence that might overstress the ability of an animal to adapt and increase the risk of metabolic disorders, it should be evident that the main causalities do not occur on the cell or genome level and thus cannot be solved there. In the first place, difficulties to adapt are due to extended and dynamic gaps between nutrient requirements and nutrient supply that already begin with the nutrient partitioning on the farm or even a larger scale. The central role of the whole organism, the genome being only a minor part, is to deal with the challenges deriving from the gaps between synchronized and tight causal coupling among the subsystems on the different scales. Correspondingly, it is the organism as a whole that succeeds or fails in adapting to nutritional change and disturbance.

Biological systems as a whole embody solutions to important biological challenges. Complex systems conveniently provide a conceptual framework and effective tools to solve problems and to decouple emergent and immergent features from molecules to organisms and back. The latter is also described as downward causation meaning the necessity for all processes at the lower level of hierarchy to act in conformity to the laws of the higher level. This means that some macro-level constraints are expected to cascade back onto micro-levels, the macro-level being itself an emergence [191]. Accordingly, pathological symptoms indicating disturbed adaptation to nutritional and metabolic changes cannot be reduced to a single factor but result from the complex interactions between various processes at different scales, thus expressing the present pathophysiology state of the whole organism.

Biological systems are goal-driven and above all they strive to survive and sustain. Facing a restricted availability of resources, while being confronted with extended change and disturbance, some animals are more successful than others at coping. However, animals that show competitive advantages in a specific unbalanced situation do not necessarily have an advantage in a quite different nutritional situation and might not show better results in the long run with respect to lifetime efficiency. The fact that metabolic disorders and production diseases have been occurring for a long time might suggest looking for solutions in the long run, e.g., breeding measures. However, metabolic disorders and production diseases are an instant problem, causing suffering, and thus require shot-term reactions and counteractive measures. Metabolic problems that derive from inappropriate and unbalanced nutrient supplies cannot be solved by selections between different genotypes but can only be reduced by the farm management in the specific context and environmental conditions. There is a significant body of knowledge on how to provide optimal farm management and prevention of bovine metabolic disorders and production diseases. For instance, tremendous efforts have been made in animal nutrition to assess and predict nutrient requirements of cows in their various life stages and following different animal and environment related variables. However, knowledge on how to adapt requirements to each unique case does not provide information on how to deal with the intra- and inter-individual variability in both nutrient demands and nutrient supply within a herd, and how to support endangered animals in coping when facing a large gap between demand and supply.

In the light of high prevalences and incidences of metabolic disorders and production diseases, it is obvious that many farm systems do not provide adequate solutions to the apparent problems; they thus overstretch the adaptation capacities of their animals. Farm systems themselves are facing fierce competition and have to adapt to change and disturbance; amongst others in feed and food markets. Farmers thus often lack the resources in terms of quality feed, labor time, investments, know how, *etc.*, needed to improve the status of animal health and welfare. Partitioning of the available resources in the most efficient way is not only a challenge for the animal but also for farm management. It can be expected that farmers would be willing to invest more in animal health and welfare if they could expect a competitive advantage. A large variation between farms regarding failure and preventive costs indicate that there is room for improvement on many farms meaning an appropriate balance which benefits a farm's profitability.

The task of agricultural and animal science should be to identify the most effective preventive measures and/or therapeutic options for farm animals but also the most cost-effective and economic approach in the corresponding complex farm system. In the first place, trade-offs have to be addressed on the scale where they emerge (principle of subsidiarity), following a bi-directional approach which simultaneously reflects upward and downward causation. Applied agricultural and animal science should support the farmer in striving for the goal to find an economic balance between productivity and animal health and welfare to the benefit of the animals, the farmers, and the prosperity of the common good. This can only be obtained by the development of more comprehensive concepts and evidence based solutions which simultaneously consider a larger number of causal relationships to discover regularities beyond the scope of individual scientific disciplines.

11. CONCLUSIONS

The pathophysiology of metabolic disorders and postpartum diseases is quite complex, interrelated with many processes within or outside the intermediary metabolism, they are also multifactorial and often driven by several interconnected risk factors. In this way, it is difficult for both scientists and the farm management to grasp the complexity and make the appropriate decisions to improve the adaptation capacity of dairy cows, particularly in the transition period. Metabolic processes in early lactation are dominated by a pull effect originating from the high glucose demand of the epithelial cells in the udder and by a large variation on the various process levels. The capacity of the dairy cows to cope with unbalanced situations between requirements and supply is limited and often exhausted. Especially in the transition period, dairy cows face varying initial conditions (e.g., body condition, medical history) and different boundary conditions (e.g., feeding and housing conditions) which interact with physiological regularities and thus result in highly variable outcomes. As a result, the interconnectedness of numerous factors on different scales create a large variation in adaptation capacities and in the occurrence of metabolic disorders and production diseases both on the animal and dairy farm levels.

It is concluded that the fundamental principles of metabolism are based on network properties which are able to contribute to the adaptability of the organism and not on the properties of single components. A metabolic network is a functional union of the different organs and tissues involved, providing a responsiveness to metabolic change and disturbance as a functional unit or as a whole organism, respectively. An understanding of symbiosis implies that the inter-tissue competition becomes an insufficient explanation for metabolic function but must incorporate a wide range of very intimate modes of cooperation. In a new perspective, the object of biological investigation becomes the process that pervades multiple levels of organization. Cells, and tissues must be understood as sub-systems within the framework of organs, and these sub-systems are embedded within broader organismal systems of interconnected and communicating structures. This represents a cooperative aspect of life and carries much importance in relation to fitness and reproduction.

The ability to cope with suboptimal living conditions and a large gap between nutrient supply and demand cannot be left to the dairy cow itself. It lies in the first place under the responsibility of the farm management to regulate nutrient and energy input as well as output to prevent the exhaustion of this adaptation capacity. On the other hand, dairy farming is a business that people engage in for financial gains. Dairy production has capitalized on the metabolic drive of the dairy cow to ensure provision of milk to the calf by selecting for increasingly higher milk production. This strategy has been very successful. Unfortunately, the goals of economic efficiency and optimal dairy cow health are often in conflict. Metabolic disorders in early lactation indicate an overstressed ability of the animals to adapt to living conditions that do not appropriately provide the specific nutrient and energy requirements. Dairy cows will succeed more easily in adaptation and in avoiding dysfunctional processes in the transition period when the gap between nutrient and energy demands and supply is restricted. The fact that on average, one in two cows succumbs to health problems during the transition period highlights the fragility of many production systems. From these previous considerations, it seems obvious that animals within a herd will be better able to adapt in order to survive, if appropriate resources and living conditions are offered which meet the individual requirements at the different stages to a high degree and if nutritional and other disturbances are reduced to a minimum. Any attempts to reduce prevalences of metabolic disorders and associated production diseases should rely on a continuous and comprehensive monitoring of appropriate indicators on the farm level. If low prevalences of metabolic disorders and production diseases would go along with a competitive advantage, it would support the farm management in providing appropriate living conditions. As a first step, low prevalences for the most relevant production disease should be established as a separate production goal, continuously assessed by appropriate criteria and monitored as an emergent health output of the farm practice. Furthermore, those farms able to combine a low prevalence of production diseases with a high productive output should be rewarded by premium prices. In the light of a lack of success in a strong reduction in the prevalence of metabolic disorders and production diseases; the deficiencies in reductionist approaches on the farm level are becoming increasingly apparent, challenging the disciplines of agricultural and animal science. Thus, scientists should ask themselves whether the focus on single aspects has contributed to promoting the tendency of oversimplification and of widely ignoring the context in which adaptation processes take place. One of the central questions should be to what degree each of the subsystems contributes to the benefit of the superordinate systems and *vice versa*.

ACKNOWLEDGMENTS

The author thanks Hoischen-Taubner, K. Müller, and F. Brand for their support and the very fruitful and comprehensive discussions we shared on this issue of metabolic disorders.

CONFLICTS OF INTEREST

The author declares no conflicts of interest.

REFERENCES

1. Maturana, H.R.; Varela, F.J. Autopoiesis and Cognition: The Realization of the Living; Springer Netherlands: Dordrecht, The Netherlands, 1980.

2. Piaget, J. Biology and Knowledge: An Essay on the Relations between Organic Regulations and Cognitive Processes; Edinburgh University Press; University of Chicago Press: Edinburgh, UK, 1971.

3. Broom, D.M. Assessing the welfare of modified or treated animals. Livest. Prod. Sci. 1993, 36, 39–54.

4. Wingfield, J.C.; Romero, L.M. Adrenocortical responses to stress and their modulation in free-living vertebrates. InHandbook of Physiology, Section 7, Coping with the Environment: Neural and Endocrine Mechanisms; McEwen, B.S., Ed.; Oxford University Press: Oxford, UK, 2000; Volume 4, pp. 211–236.

5. Leroy, J.L.; Vanholder, T.; van Knegsel, A.T.M.; Garcia-Ispierto, I.; Bols, P.E.J. Nutrient prioritization in dairy cows early postpartum: Mismatch between metabolism and fertility? Reprod. Domest. Anim. 2008, 43 (Suppl. 2), 96–103.

6. Hogeveen, H. Costs of production diseases. In Proceedings of the 27th World Buiatrics Congress, Cairns, Australia, 27 July–1 August 2014; pp. 36–42.

7. Drackley, J.K. Biology of dairy cows during the transition period: The final frontier? J. Dairy Sci. 1999, 82, 2259–2273.

8. Bell, A.W. Regulation of organic nutrient metabolism during transition from late pregnancy to early lactation. J. Anim. Sci. 1995, 73, 2804–2819.

9. Sordillo, L.M.; Raphael, W. Significance of metabolic stress, lipid mobilization, and inflammation on transition cow disorders. Vet. Clin. North Am. Food Anim. Pract. 2013, 29, 267–278.

10. Ametaj, B.N. Metabolic disorders of dairy cattle. In Encyclopedia of Life Support Systems (EOLSS): Veterinary Science; Developed Under the Auspices of the UNESCO, Eolss Publishers: Paris, France, 2014.

11. Mulligan, F.J.; Doherty, M.L. Production diseases of the transition cow. Vet. J. 2008, 176, 3–9.

12. Oetzel, G.R. Undertaking nutritional diagnostic investigations. Vet. Clin. North Am. Food Anim. Pract. 2014, 30, 765–788.

13. Herdt, T.H. Metabolic diseases of dairy cattle. Vet. Clin. North Am. Food Anim. Pract. 2013, 29.

14. Brand, F.S.; Jax, K. Focusing the meaning(s) of resilience: Resilience as a descriptive concept and a boundary object.Ecol. Soc. 2007, 12, 23.

15. Darnhofer, I. Strategies of family farms to strengthen their resilience. Environ. Policy Gov. 2010, 20, 212–222.

16. Di Paolo, E. Autopoiesis, adaptivity, teleology, agency. Phenomenol. Cogn. Sci. 2005, 4, 429–452.

17. Saborido, C.; Moreno, A. Biological pathology from an organizational perspective. Theor. Med. Bioeth. 2015, 36, 83–95.

18. Wingfield, J.C. Allostatic load and life cycles: Implications for neuroendocrine control mechanisms. In Allostasis, Homeostasis, and the Costs of Physiological Adaptation; Schulkin, J., Ed.; Cambridge University Press: Cambridge, UK, 2004; pp. 302–342.

19. Montévil, M.; Mossio, M. Biological organisation as closure of constraints. J. Theor. Biol. 2015, 372, 179–191.

20. Bernard, C. Introduction à L'étude de la Médecine Expérimentale; Baillière: Paris, France, 1865.

21. Bernard, C. Leçons sur les Phénomènes de la vie Communs aux Animaux et aux Végétaux; J.-B. Baillière et fils: Paris, France, 1878.

22. Cannon, W.B. Organization for physiological homeostasis. Physiol. Rev. 1929, 9, 399–431.

23. Ashby, W.R. An Introduction to Cybernetics, 2nd ed.; Chapman & Hall: London, UK, 1956.

24. Wiener, N. Cybernetics: Or Control and Communication in the Animal and the Machine, 2nd ed.; M.I.T. Press: Cambridge, MA, USA, 1948.

25. Cannon, W.B. The Wisdom of the Body; WW Norton & Co.: New York, NY, USA, 1932.

26. Selye, H. Homeostasis and heterostasis. Perspect. Biol. Med. 1973, 16, 441–445.

27. Waddington, C.H. The basic ideas of biology. In Towards a Theoretical Biology: Prolegomena; Waddington, C.H., Ed.; Edinburgh University Press: Edinburgh, UK; Aldine Publ. Co.: Chicago, IL, USA, 1968; pp. 238–253.

28. Letelier, J.-C.; Cárdenas, M.L.; Cornish-Bowden, A. From l'homme machine to metabolic closure: Steps towards understanding life. J. Theor. Biol. 2011, 286, 100–113.

29. Varela, F.G.; Maturana, H.R.; Uribe, R. Autopoiesis: The organization of living systems, its characterization and a model. Biosystems 1974, 5, 187–196.

30. Bauman, D.E.; Currie, B.W. Partitioning of nutrients during pregnancy and lactation: A review of mechanisms involving homeostasis and homeorhesis. J. Dairy Sci. 1980, 63, 1514–1529.

31. Mrosovsky, N. Rheostasis: The Physiology of Change; Oxford University Press: New York, NY, USA, 1990.

32. Knight, C.H.; Beever, D.E.; Sorensen, A. Metabolic loads to be expected from different genotypes under different systems: Metabolic stress in dairy cows. Br. Soc. Anim. Sci. Occas. Publ. 1999, 27–36.

33. Sterling, P.; Eyer, J. Allostasis: A new paradigm to explain arousal pathology. In Handbook of Life Stress, Cognition and Health; Fisher, S., Reason, J., Eds.; John Wiley & Sons: Hoboken, NJ, USA, 1988; pp. 629–649.

34. McEwen, B.S. Stress, adaptation, and disease: Allostasis and allostatic load. Ann. N. Y. Acad. Sci. 1998, 840, 33–44.

35. McEwen, B.S.; Wingfield, J.C. The concept of allostasis in biology and biomedicine. Horm. Behav. 2003, 43, 2–15.

36. Herdt, T.H. The history and influence of the ICPD. In Production Diseases in Farm Animals: 12th International Conference; Joshi, N.P., Herdt, T.H., Eds.; Wageningen Academic Publishers: Wageningen, The Netherlands, 2006; pp. 19–22.

37. Knaus, W. Dairy cows trapped between performance demands and adaptability. J. Sci. Food Agric. 2009, 89, 1107–1114.

38. Rauw, W.; Kanis, E.; Noordhuizen-Stassen, E.; Grommers, F. Undesirable side effects of selection for high production efficiency in farm animals: A review. Livest. Prod. Sci. 1998, 56, 15–33.

39. LeBlanc, S. Managing critical periods—transition dairy cows. In Book of Abstracts, 15th Conference on Production Diseases in Farm Animals; Dalin, G., Ed.; Swedish University of Agricultural Science (SLU): Uppsala, Sweden, 2013; pp. 62–65.

40. Van Saun, R.J.; Sniffen, C.J. Transition cow nutrition and feeding management for disease prevention. Vet. Clin. North Am. Food Anim. Pract. 2014, 30, 689–719.

41. Dohoo, I.R.; Wayne Martin, S. Disease, production and culling in Holstein-Friesian cows: IV. Effects of disease on production. Prev. Vet. Med. 1984, 2, 755–770.

42. Andersson, L.; Emanuelson, U. An epidemiological study of hyperketonaemia in Swedish dairy cows; Determinants and the relation to fertility. Prev. Vet. Med. 1985, 3, 449–462.

43. Erb, H.N. Interrelationships among production and clinical disease in dairy cattle: A review. Can. Vet. J. 1987, 28, 326–329.

44. Melendez, P.; Marin, M.P.; Robles, J.; Rios, C.; Duchens, M.; Archbald, L. Relationship between serum nonesterified fatty acids at calving and the incidence of periparturient diseases in Holstein dairy cows. Theriogenology 2009, 72, 826–833.

45. Ospina, P.A.; Nydam, D.V.; Stokol, T.; Overton, T.R. Evaluation of nonesterified fatty acids and beta-hydroxybutyrate in transition dairy cattle in the northeastern United States: Critical thresholds for prediction of clinical diseases. J. Dairy Sci. 2010, 93, 546–554.

46. Goff, J.P.; Horst, R.L. Physiological changes at parturition and their relationship to metabolic disorders. J. Dairy Sci.1997, 80, 1260–1268.

47. Drackley, J.K.; Dann, H.M.; Douglas, G.N.; Guretzky, N.A.J.; Litherland, N.B.; Underwood, J.P.; Loor, J.J. Physiological and pathological adaptations in dairy cows that may increase susceptibility to periparturient diseases and disorders. Italian J. Animal Sci. 2005.

48. Van der Drift, S.G.A.; Jorritsma, R.; Schonewille, J.T.; Knijn, H.M.; Stegeman, J.A. Routine detection of hyperketonemia in dairy cows using Fourier transform infrared spectroscopy analysis of β-hydroxybutyrate and acetone in milk in combination with test-day information. J. Dairy Sci. 2012, 95, 4886–4898.

49. Kessel, S.; Stroehl, M.; Meyer, H.H.D.; Hiss, S.; Sauerwein, H.; Schwarz, F.J.; Bruckmaier, R.M. Individual variability in physiological adaptation to metabolic stress during early lactation in dairy cows kept under equal conditions. J. Anim. Sci. 2008, 86, 2903–2912.

50. Aitken, S.; Corl, C.; Sordillo, L. Immunopathology of mastitis: Insights into disease recognition and resolution: Journal of Mammary Gland Biology and Neoplasia. J. Mammary Gland Biol. Neoplasia 2011, 16, 291–304.

51. Esposito, G.; Irons, P.C.; Webb, E.C.; Chapwanya, A. Interactions between negative energy balance, metabolic diseases, uterine health and immune response in transition dairy cows. Anim. Reprod. Sci. 2014, 144, 60–71.

52. Loiselle, M.C.; Ster, C.; Talbot, B.G.; Zhao, X.; Wagner, G.F.; Boisclair, Y.R.; Lacasse, P. Impact of postpartum milking frequency on the immune system and the blood metabolite concentration of dairy cows. J. Dairy Sci. 2009, 92, 1900–1912.

53. Wathes, D.C.; Cheng, Z.; Chowdhury, W.; Fenwick, M.A.; Fitzpatrick, R.; Morris, D.G.; Patton, J.; Murphy, J.J. Negative energy balance alters global gene expression and immune responses in the uterus of postpartum dairy cows. Physiol. Genomics 2009, 39, 1–13.

54. Moyes, K.M.; Drackley, J.K.; Morin, D.E.; Rodriguez-Zas, S.L.; Everts, R.E.; Lewin, H.A.; Loor, J.J. Mammary gene expression profiles during an intramammary challenge reveal potential mechanisms linking negative energy balance with impaired immune response. Physiol. Genomics 2010, 41, 161–170.

55. Burvenich, C.; Bannerman, D.D.; Lippolis, J.D.; Peelman, L.; Nonnecke, B.J.; Kehrli, M.E.; Paape, M.J. Cumulative physiological events influence the inflammatory response of the bovine udder to Escherichia coli infections during the transition period. J. Dairy Sci. 2007, 90 (Suppl. 1), E39–E54.

56. LeBlanc, S.J.; Osawa, T.; Dubuc, J. Reproductive tract defense and disease in postpartum dairy cows. Theriogenology2011, 76, 1610–1618.

57. Sheldon, I.M.; Cronin, J.; Goetze, L.; Donofrio, G.; Schuberth, H.-J. Defining postpartum uterine disease and the mechanisms of infection and immunity in the female reproductive tract in cattle. Biol. Reprod. 2009, 81, 1025–1032.

58. Huzzey, J.M.; Veira, D.M.; Weary, D.M.; von Keyserlingk, M.A.G. Prepartum behavior and dry matter intake identify dairy cows at risk for metritis. J. Dairy Sci. 2007, 90, 3220–3233.

59. Ster, C.; Loiselle, M.-C.; Lacasse, P. Effect of postcalving serum nonesterified fatty acids concentration on the functionality of bovine immune cells. J. Dairy Sci. 2012, 95, 708–717.

60. Hammon, D.S.; Evjen, I.M.; Dhiman, T.R.; Goff, J.P.; Walters, J.L. Neutrophil function and energy status in Holstein cows with uterine health disorders. Vet. Immunol. Immunopathol. 2006, 113, 21–29.

61. Sordillo, L.M.; Contreras, G.A.; Aitken, S.L. Metabolic factors affecting the inflammatory response of periparturient dairy cows. Anim. Health Res. Rev. 2009, 10, 53–63.

62. Sordillo, L.M.; Mavangira, V. The nexus between nutrient metabolism, oxidative stress and inflammation in transition cows. Anim. Prod. Sci. 2014, 54, 1204–1214.

63. Farney, J.K.; Mamedova, L.K.; Coetzee, J.F.; KuKanich, B.; Sordillo, L.M.; Stoakes, S.K.; Minton, J.E.; Hollis, L.C.; Bradford, B.J. Anti-inflammatory salicylate treatment alters the metabolic adaptations to lactation in dairy cattle. Am. J. Physiol. Regul. Integr. Comp. Physiol. 2013, 305, R110–R117.

64. Huzzey, J.M.; Mann, S.; Nydam, D.V.; Grant, R.J.; Overton, T.R. Associations of peripartum markers of stress and inflammation with milk yield and reproductive performance in Holstein dairy cows. Prev. Vet. Med. 2015, 120, 291–297.

65. Abuelo, A.; Hernández, J.; Benedito, J.L.; Castillo, C. The importance of the oxidative status of dairy cattle in the periparturient period: Revisiting antioxidant supplementation. J. Anim. Physiol. Anim. Nutr. (Berl.) 2015.

66. Taylor, V.; Beever, D.; Wathes, D.; Kebreab, E.; Mills, J. Physiological adaptions to milk production that affect the fertility of high yielding dairy cows. In Dairying: Using Science to Meet Consumers' Needs; University of Reading: Reading, UK, 2004; pp. 37–71.

67. Beever, D.E. The impact of controlled nutrition during the dry period on dairy cow health, fertility and performance.Anim. Reprod. Sci. 2006, 96, 212–226.

68. Robinson, J.J.; Ashworth, C.J.; Rooke, J.A.; Mitchell, L.M.; McEvoy, T.G. Nutrition and fertility in ruminant livestock.Anim. Feed Sci. Technol. 2006, 126, 259–276.

69. Collard, B.L.; Boettcher, P.J.; Dekkers, J.; Petitclerc, D.; Schaeffer, L.R. Relationships between energy balance and health traits of dairy cattle in early lactation. J. Dairy Sci. 2000, 83, 2683–2690.

70. Razeto-Barry, P. Autopoiesis 40 years later. A review and a reformulation. Orig. Life Evol. Biosph. 2012, 42, 543–567.

71. Grant, R.J.; Albright, J.L. Feeding behavior and management factors during the transition period in dairy cattle. J. Anim. Sci. 1995, 73, 2791–2803.

72. Tylutki, T.P.; Fox, D.G.; Durbal, V.M.; Tedeschi, L.O.; Russell, J.B.; van Amburgh, M.E.; Overton, T.R.; Chase, L.E.; Pell, A.N. Cornell Net Carbohydrate and Protein System: A model for precision feeding of dairy cattle. Anim. Feed Sci. Technol. 2008, 143, 174–202.

73. Allen, M.S.; Piantoni, P. Metabolic control of feed intake: Implications for metabolic disease of fresh cows. Vet. Clin. North Am. Food Anim. Pract. 2013, 29, 279–297.

74. Allen, M.S. Effects of diet on short-term regulation of feed intake by lactating dairy cattle. J. Dairy Sci. 2000, 83, 1598–1624.

75. Allen, M.S.; Bradford, B.J.; Oba, M. The hepatic oxidation theory of the control of feed intake and its application to ruminants. J. Anim. Sci. 2009, 87, 3317–3334.

76. Meyer, U.; Horstmann, K.; Kaske, M.; Flachowsky, G. Inter- and Intra-Individual Variation of Feed Intake and Metabolic Parameters of Dairy Cows Related to Energy Supply; European Association for Animal Production: Roma, Italy, 2007; Volume 124, p. 571.

77. Dann, H.M. Dietary Energy Restriction during Late Gestation in Multiparous Holstein Cows. Ph.D. Thesis, University of Illinois, Champaign, IL, USA, 2004.

78. Lean, I.J.; Golder, H.M.; Hall, M.B. Feeding, evaluating, and controlling rumen function. Vet. Clin. North Am. Food Anim. Pract. 2014, 30, 539–575.

79. Proudfoot, K.L.; Veira, D.M.; Weary, D.M.; von Keyserlingk, M.A.G. Competition at the feed bunk changes the feeding, standing, and social behavior of transition dairy cows. J. Dairy Sci. 2009, 92, 3116–3123.

80. Cook, N.B.; Bennett, T.B.; Nordlund, K.V. Effect of free stall surface on daily activity patterns in dairy cows with relevance to lameness prevalence. J. Dairy Sci. 2004, 87, 2912–2922.

81. Kuhla, B. Central Regulation of Feed Intake in Early Lactation; Leibniz Institute: Jena, Germany, 2012.

82. Merchen, N.R.; Elizalde, J.C.; Drackley, J.K. Current perspective on assessing site of digestion in ruminants. J. Anim. Sci. 1997, 75, 2223–2234.

83. Kamra, D.N. Rumen microbial ecosystem. Curr. Sci. 2005, 89, 124–135.

84. Tajima, K.; Aminov, R.I.; Nagamine, T.; Matsui, H.; Nakamura, M.; Benno, Y. Diet-dependent shifts in the bacterial population of the rumen

revealed with real-time PCR. Appl. Environ. Microbiol. 2001, 67, 2766–2774.

85. Brown, M.S.; Krehbiel, C.R.; Galyean, M.L.; Remmenga, M.D.; Peters, J.P.; Hibbard, B.; Robinson, J.; Moseley, W.M. Evaluation of models of acute and subacute acidosis on dry matter intake, ruminal fermentation, blood chemistry, and endocrine profiles of beef steers. J. Anim. Sci. 2000, 78, 3155–3168.

86. Pers-Kamczyc, E.; Zmora, P.; Cieślak, A.; Szumacher-Strabel, M. Development of nucleic acid based techniques and possibilities of their application to rumen microbial ecology research. J. Anim. Feed Sci. 2011, 20, 315–337.

87. Stewart, C.S.; Flint, H.J.; Bryant, M.P. The rumen bacteria. In The Rumen Microbial Ecosystem; Hobson, P.N., Stewart, C.S., Eds.; Springer: Dordrecht, The Netherlands, 1997; pp. 10–72.

88. Dijkstra, J.; Kebreab, E.; Mills, J.A.N.; Pellikaan, W.F.; López, S.; Bannink, A.; France, J. Predicting the profile of nutrients available for absorption: From nutrient requirement to animal response and environmental impact. Animal2007, 1, 99–111.

89. Nozière, P.; Hoch, T. Modelling fluxes of volatile fatty acids from rumen to portal blood. In Nutrient Digestion and Utilization in Farm Animals: Modelling Approaches; Kebreab, E., Nozière, P., Hoch, T., Eds.; CABI Publishing: Wallingford, CT, USA, 2006; pp. 40–47.

90. Dijkstra, J.; Boer, H.; van Bruchem, J.; Bruining, M.; Tamminga, S. Absorption of volatile fatty acids from the rumen of lactating dairy cows as influenced by volatile fatty acid concentration, pH and rumen liquid volume. Br. J. Nutr. 1993,69, 385–396.

91. Rabelo, E.; Rezende, R.L.; Bertics, S.J.; Grummer, R.R. Effects of transition diets varying in dietary energy density on lactation performance and ruminal parameters of dairy cows. J. Dairy Sci. 2003, 86, 916–925.

92. Dijkstra, J.; Ellis, J.L.; Kebreab, E.; Strathe, A.B.; López, S.; France, J.; Bannink, A. Ruminal pH regulation and nutritional consequences of low pH. Anim. Feed Sci. Technol. 2012, 172, 22–33.

93. Stein, S.K.; Sundrum, A. Detection of SARA in the transition period of dairy cows. J. Dairy Sci. 2015. submitted.

94. Bossen, D.; Mertens, D.R.; Weisbjerg, M.R. Influence of fermentation methods on neutral detergent fiber degradation parameters. J. Dairy Sci. 2008, 91, 1464–1476.

95. Hall, M.B. Dietary starch source and protein degradability in diets containing sucrose: Effects on ruminal measures and proposed mechanism for degradable protein effects. J. Dairy Sci. 2013, 96, 7093–7109.

96. Argyle, J.L.; Baldwin, R.L. Effects of amino acids and peptides on rumen microbial growth yields. J. Dairy Sci. 1989,72, 2017–2027.

97. Oba, M.; Allen, M.S. Effects of corn grain conservation method on feeding behavior and productivity of lactating dairy cows at two dietary starch concentrations. J. Dairy Sci. 2003, 86, 174–183.

98. Firkins, J.L.; Allen, M.S.; Oldick, B.S.; St-Pierre, N.R. Modeling ruminal digestibility of carbohydrates and microbial protein flow to the duodenum. J. Dairy Sci. 1998, 81, 3350–3369.

99. Erdman, R.A.; Varner, M. Fixed yield responses to increased milking frequency. J. Dairy Sci. 1995, 78, 1199–1203.

100. Stefanon, B.; Colitti, M.; Gabai, G.; Knight, C.H.; Wilde, C.J. Mammary apoptosis and lactation persistency in dairy animals. J. Dairy Res. 2002, 69, 37–52.

101. Murney, R.; Stelwagen, K.; Wheeler, T.T.; Margerison, J.K.; Singh, K. The effects of milking frequency in early lactation on milk yield, mammary cell turnover, and secretory activity in grazing dairy cows. J. Dairy Sci. 2015, 98, 305–311.

102. Bauman, D.E.; Mather, I.H.; Wall, R.J.; Lock, A.L. Major advances associated with the biosynthesis of milk. J. Dairy Sci.2006, 89, 1235–1243.

103. Gardner, N.H.; Reynolds, C.K.; Phipps, R.H.; Jones, A.K.; Beever, D.E. Effects of different diet supplements in the pre- and post-partum period on reproductive performance in the dairy cow. In Fertility in the High Producing Dairy Cow; Occasional Publication No. 26; British Society of Animal Science: London, UK, 2001; pp. 313–322.

104. Van Knegsel, A.; van den Brand, H.; Dijkstra, J.; van Straalen, W.M.; Jorritsma, R.; Tamminga, S.; Kemp, B. Effect of glucogenic vs. lipogenic diets on energy balance, blood metabolites, and reproduction in primiparous and multiparous dairy cows in early lactation. J. Dairy Sci. 2007, 90, 3397–3409.

105. Hattan, A.J. Energy Utilisation in High Yielding Dairy Cows; University of Reading: Reading, UK, 2003.

106. De Vries, M.J.; van der Beek, S.; Kaal-Lansbergen, L.; Ouweltjes, W.; Wilmink, J.B. Modeling of energy balance in early lactation and the

effect of energy deficits in early lactation on first detected estrus postpartum in dairy cows. J. Dairy Sci. 1999, 82, 1927–1934.

107. Eastridge, M.L. Major advances in applied dairy cattle nutrition. J. Dairy Sci. 2006, 89, 1311–1323.

108. Van Arendonk, J.A.; Nieuwhof, G.J.; Vos, H.; Korver, S. Genetic aspects of feed intake and efficiency in lactating dairy heifers. Livest. Prod. Sci. 1991, 29, 263–275.

109. Beerda, B.; Ouweltjes, W.; Šebek, L.B.; Windig, J.J.; Veerkamp, R.F. Effects of genotype by environment interactions on milk yield, energy balance, and protein balance. J. Dairy Sci. 2007, 90, 219–228.

110. Rhoads, M.L.; Rhoads, R.P.; VanBaale, M.J.; Collier, R.J.; Sanders, S.R.; Weber, W.J.; Crooker, B.A.; Baumgard, L.H. Effects of heat stress and plane of nutrition on lactating Holstein cows: I. Production, metabolism, and aspects of circulating somatotropin. J. Dairy Sci. 2009, 92, 1986–1997.

111. Steffens, A.B.; de Boer, S.F. Impact of stress on animal intermediate metabolism. In Stress in Physiology of Animals; Balm, P.H., Ed.; Sheffield Academic Press: Sheffield, UK, 1999; pp. 108–129.

112. Ingvartsen, K.L. Feeding- and management-related diseases in the transition cow: Physiological adaptations around calving and strategies to reduce feeding-related diseases. Anim. Feed Sci. Technol. 2006, 126, 175–213.

113. Veerkamp, R.F.; Brotherstone, S. Genetic correlations between linear type traits, food intake, live weight and condition score in Holstein Friesian dairy cattle. Anim. Sci. 1997, 64, 385–392.

114. Sutter, F.; Beever, D.E. Energy and nitrogen metabolism in Holstein-Friesian cows during early lactation. Anim. Sci.2000, 70, 503–514.

115. Friggens, N.C.; Brun-Lafleur, L.; Faverdin, P.; Sauvant, D.; Martin, O. Advances in predicting nutrient partitioning in the dairy cow: Recognizing the central role of genotype and its expression through time. Animal 2013, 7 (Suppl. 1), 89–101.

116. Patton, J.; Kenny, D.A.; Mee, J.F.; O'Mara, F.P.; Wathes, D.C.; Cook, M.; Murphy, J.J. Effect of milking frequency and diet on milk production, energy balance, and reproduction in dairy cows. J. Dairy Sci. 2006, 89, 1478–1487.

117. Bradford, B.J. Immunity and inflammation in transition cows. In Proceedings of AABP Annual Meeting, St. Louis, MO, USA, 22–24 September 2011; pp. 48–52.

118. Boutinaud, M.; Ben Chedly, M.H.; Delamaire, E.; Guinard-Flament, J. Milking and feed restriction regulate transcripts of mammary epithelial cells purified from milk. J. Dairy Sci. 2008, 91, 988–998.

119. Bell, A.W.; Burhans, W.S.; Overton, T.R. Protein nutrition in late pregnancy, maternal protein reserves and lactation performance in dairy cows. Proc. Nutr. Soc. 2000, 59, 119–126.

120. Lucy, M.C. Mechanisms linking the somatotrophic axis with insulin: Lesson from the postpartum dairy cow. Proc. N. Z. Soc. Anim. Prod. 2004, 64, 19–23.

121. Vernon, R.G. Nutrient partitioning, lipid metabolism and relevant imbalances. In Recent Developments and Perspectives in Bovine Medicine; Kasle, M., Scholz, H., Holtershinken, M., Eds.; Clinik fur Rinderkrankheiten: Hanover, Germany, 2002.

122. Zhao, F.-Q.; Dixon, W.T.; Kennelly, J.J. Localization and gene expression of glucose transporters in bovine mammary gland. Comp. Biochem. Physiol. B Biochem. Mol. Biol. 1996, 115, 127–134.

123. Butler, S.T.; Marr, A.L.; Pelton, S.H.; Radcliff, R.P.; Lucy, M.C.; Butler, W.R. Insulin restores GH responsiveness during lactation-induced negative energy balance in dairy cattle: Effects on expression of IGF-I and GH receptor 1A.J. Endocrinol. 2003, 176, 205–217.

124. Lucy, M.C. Fertility in high-producing dairy cows: Reasons for decline and corrective strategies for sustainable improvement. Soc. Reprod. Fertil. Suppl. 2007, 64, 237–254.

125. Osborn, O.; Olefsky, J.M. The cellular and signaling networks linking the immune system and metabolism in disease.Nat. Med. 2012, 18, 363–374.

126. McNamara, J.P. Regulation of adipose tissue metabolism in support of lactation. J. Dairy Sci. 1991, 74, 706–719.

127. McNamara, J.P.; Harrison, J.H.; Kincaid, R.L.; Waltner, S.S. Lipid metabolism in adipose tissue of cows fed high fat diets during lactation. J. Dairy Sci. 1995, 78, 2782–2796.

128. Smith, T.H.; McNamara, J.P. Regulation of bovine adipose tissue metabolism during lactation. 6. Cellularity and hormone-sensitive lipase activity as affected by genetic merit and energy intake. J. Dairy Sci. 1990, 73, 772–783.

129. Lucy, M.C.; Jiang, H.; Kobayashi, Y. Changes in the somatotrophic axis associated with the initiation of lactation. J. Dairy Sci. 2001, 84, E113–E119.

130. Rukkwamsuk, T.; Wensing, T.; Geelen, M.J. Effect of overfeeding during the dry period on regulation of adipose tissue metabolism in dairy cows during the periparturient period. J. Dairy Sci. 1998, 81, 2904–2911.

131. Rukkwamsuk, T.; Wensing, T.; Geelen, M.J. Effect of overfeeding during the dry period on the rate of esterification in adipose tissue of dairy cows during the periparturient period. J. Dairy Sci. 1999, 82, 1164–1169.

132. Locher, L.; Häussler, S.; Laubenthal, L.; Singh, S.P.; Winkler, J.; Kinoshita, A.; Kenéz, Á.; Rehage, J.; Huber, K.; Sauerwein, H. Effect of increasing body condition on key regulators of fat metabolism in subcutaneous adipose tissue depot and circulation of non-lactating dairy cows. J. Dairy Sci. 2015, 98, 1057–1068.

133. Katamoto, H.; Yukawa, T.; Shimada, Y. Lipogenic and lipolytic activities in isolated adipocytes from cattle with fat necrosis. Res. Vet. Sci. 1996, 61, 214–217.

134. Mann, S.; Yepes, F.A.L.; Overton, T.R.; Wakshlag, J.J.; Lock, A.L.; Ryan, C.M.; Nydam, D.V. Dry period plane of energy: Effects on feed intake, energy balance, milk production, and composition in transition dairy cows. J. Dairy Sci.2015, 98, 3366–3382.

135. Andersen, J.B.; Ridder, C.; Larsen, T. Priming the cow for mobilization in the periparturient period: Effects of supplementing the dry cow with saturated fat or linseed. J. Dairy Sci. 2008, 91, 1029–1043.

136. Litherland, N.B.; da Silva, D.N.L.; Hansen, W.P.; Davis, L.; Emanuele, S.; Blalock, H. Effects of prepartum controlled-energy wheat straw and grass hay diets supplemented with starch or sugar on periparturient dairy cow performance and lipid metabolism. J. Dairy Sci. 2013, 96, 3050–3063.

137. Newman, A.; Mann, S.; Nydam, D.V.; Overton, T.R.; Behling-Kelly, E. Impact of dietary plane of energy during the dry period on lipoprotein parameters in the transition period in dairy cattle. J. Anim. Physiol. Anim. Nutr. (Berl.) 2015.

138. Overton, T.R.; Drackley, J.K.; Ottemann-Abbamonte, C.J.; Beaulieu, A.D.; Emmert, L.S.; Clark, J.H. Substrate utilization for hepatic gluconeogenesis is altered by increased glucose demand in ruminants. J. Anim. Sci. 1999, 77, 1940–1951.

139. Kreipe, L.; Vernay, M.C.M.B.; Oppliger, A.; Wellnitz, O.; Bruckmaier, R.M.; van Dorland, H.A. Induced hypoglycemia for 48

hours indicates differential glucose and insulin effects on liver metabolism in dairy cows. J. Dairy Sci. 2011, 94, 5435–5448.

140. Reynolds, C.K.; Aikman, P.C.; Lupoli, B.; Humphries, D.J.; Beever, D.E. Splanchnic metabolism of dairy cows during the transition from late gestation through early lactation. J. Dairy Sci. 2003, 86, 1201–1217.

141. Herdt, T.H. Ruminant adaptation to negative energy balance. Influences on the etiology of ketosis and fatty liver. Vet. Clin. North Am. Food Anim. Pract. 2000, 16, 215–230.

142. Hammon, H.M.; Stürmer, G.; Schneider, F.; Tuchscherer, A.; Blum, H.; Engelhard, T.; Genzel, A.; Staufenbiel, R.; Kanitz, W. Performance and metabolic and endocrine changes with emphasis on glucose metabolism in high-yielding dairy cows with high and low fat content in liver after calving. J. Dairy Sci. 2009, 92, 1554–1566.

143. Cadórniga-Valiño, C.; Grummer, R.R.; Armentano, L.E.; Donkin, S.S.; Bertics, S.J. Effects of fatty acids and hormones on fatty acid metabolism and gluconeogenesis in bovine hepatocytes. J. Dairy Sci. 1997, 80, 646–656.

144. Strang, B.D.; Bertics, S.J.; Grummer, R.R.; Armentano, L.E. Effect of long-chain fatty acids on triglyceride accumulation, gluconeogenesis, and ureagenesis in bovine hepatocytes. J. Dairy Sci. 1998, 81, 728–739.

145. Vazquez-Añon, M.; Bertics, S.; Luck, M.; Grummer, R.R.; Pinheiro, J. Peripartum liver triglyceride and plasma metabolites in dairy cows. J. Dairy Sci. 1994, 77, 1521–1528.

146. Cornish-Bowden, A.; Cárdenas, M.L.; Letelier, J.-C.; Soto-Andrade, J.; Abarzúa, F.G. Understanding the parts in terms of the whole. Biol. Cell 2004, 96, 713–717.

147. Bannink, A.; France, J.; Lopez, S.; Gerrits, W.; Kebreab, E.; Tamminga, S.; Dijkstra, J. Modelling the implications of feeding strategy on rumen fermentation and functioning of the rumen wall. Anim. Feed Sci. Technol. 2008, 143, 3–26.

148. Sudakov, K. The theory of functional systems: General postulates and principles of dynamic organization. Integr. Physiol. Behav. Sci. 1997, 32, 392–414.

149. Oppenheim, P.; Putnam, H. The unity of science as a working hypothesis. In Concepts, Theories, and the Mind-Body Problem; Feigl, H., Ed.; University of Minnesota Press: Minneapolis, MI, USA, 1958; Volume 2, pp. 3–36.

150. Powell, A.; Dupré, J. From molecules to systems: The importance of looking both ways. Stud. Hist. Philos. Biol. Biomed. Sci. 2009, 40, 54–64.

151. Conway, G.R. The properties of agroecosystems. Agric. Syst. 1987, 24, 95–117.

152. Varela, F.J. Patterns of life: Intertwining identity and cognition. Brain Cogn. 1997, 34, 72–87.

153. Thompson, P.B. The varieties of sustainability in livestock farming. EAAP Publ. 1997, 89, 5–15.

154. Fox, C.J.; Hammerman, P.S.; Thompson, C.B. Fuel feeds function: Energy metabolism and the T-cell response. Nat. Rev. Immunol. 2005, 5, 844–852.

155. Wagner, A. Robustness and Evolvability in Living Systems; Princeton University Press: Princeton, NJ, USA, 2005.

156. Lucy, M.C. Mechanisms linking nutrition and reproduction in postpartum cows. Reprod. Suppl. 2003, 61, 415–427.

157. Buckley, F.; Dillon, P.; Rath, M.; Veerkamp, R.F. The relationship between genetic merit for yield and live weight, condition score, and energy balance of spring calving Holstein Friesian dairy cows on grass based systems of milk production. J. Dairy Sci. 2000, 83, 1878–1886.

158. Krause, K.M.; Oetzel, G.R. Understanding and preventing subacute ruminal acidosis in dairy herds: A review. Anim. Feed Sci. Technol. 2006, 126, 215–236.

159. Haerle, C.; Sundrum, A. Animal health on the farm level 2. Communication: nutrient supply on bavarian dairy farms.Züchtungskunde 2013, 85, 396–412.

160. Holtenius, K.; Agenäs, S.; Delavaud, C.; Chilliard, Y. Effects of feeding intensity during the dry period. 2. Metabolic and hormonal responses. J. Dairy Sci. 2003, 86, 883–891.

161. Rastani, R.R.; Grummer, R.R.; Bertics, S.J.; Gümen, A.; Wiltbank, M.C.; Mashek, D.G.; Schwab, M.C. Reducing dry period length to simplify feeding transition cows: Milk production, energy balance, and metabolic profiles. J. Dairy Sci. 2005, 88, 1004–1014.

162. Gümen, A.; Rastani, R.R.; Grummer, R.R.; Wiltbank, M.C. Reduced dry periods and varying prepartum diets alter postpartum ovulation and reproductive measures. J. Dairy Sci. 2005, 88, 2401–2411.

163. Van Knegsel, A.; van der Drift, S.G.A.; Cermáková, J.; Kemp, B. Effects of shortening the dry period of dairy cows on milk production, energy balance, health, and fertility: A systematic review. Vet. J. 2013, 198, 707–713.

164. Chen, J.; Gross, J.J.; van Dorland, H.A.; Remmelink, G.J.; Bruckmaier, R.M.; Kemp, B.; van Knegsel, A.T.M. Effects of dry period length and dietary energy source on metabolic status and hepatic gene expression of dairy cows in early lactation. J. Dairy Sci. 2015, 98, 1033–1045.

165. McNamara, S.; Murphy, J.J.; O'Mara, F.P.; Rath, M.; Mee, J.F. Effect of milking frequency in early lactation on energy metabolism, milk production and reproductive performance of dairy cows. Livest. Sci. 2008, 117, 70–78.

166. Phyn, C.; Kay, J.K.; Rius, A.G.; Morgan, S.R.; Roach, C.S.; Grala, T.M.; Roche, J.R. Review: Impact of short—Term alterations to milking frequency in early lactation. In Proceedings of 4th Australian Dairy Science Symposium; Caxton Press: Caldwell, NJ, USA, 2010.

167. Clark, D.A.; Phyn, C.; Tong, M.J.; Collis, S.J.; Dalley, D.E. A systems comparison of once- versus twice-daily milking of pastured dairy cows. J. Dairy Sci. 2006, 89, 1854–1862.

168. Stelwagen, K.; Phyn, C.V.C.; Davis, S.R.; Guinard-Flament, J.; Pomiès, D.; Roche, J.R.; Kay, J.K. Invited review: Reduced milking frequency: Milk production and management implications. J. Dairy Sci. 2013, 96, 3401–3413.

169. Carbonneau, E.; de Passillé, A.M.; Rushen, J.; Talbot, B.G.; Lacasse, P. The effect of incomplete milking or nursing on milk production, blood metabolites, and immune functions of dairy cows. J. Dairy Sci. 2012, 95, 6503–6512.

170. Soberon, F.; Ryan, C.M.; Nydam, D.V.; Galton, D.M.; Overton, T.R. The effects of increased milking frequency during early lactation on milk yield and milk composition on commercial dairy farms. J. Dairy Sci. 2011, 94, 4398–4405.

171. Oltenacu, P.A.; Algers, B. Selection for increased production and the welfare of dairy cows: Are new breeding goals needed? Ambio 2005, 34, 311–315.

172. Gordon, F.J.; Patterson, D.C.; Yan, T.; Porter, M.G.; Mayne, C.S.; Unsworth, E.F. The influence of genetic index for milk production on the response to complete diet feeding and the utilization of energy and nitrogen. Anim. Sci. 1995,61, 199–210.

173. Tyrrell, H.F.; Brown, A.C.; Reynolds, P.J.; Haaland, G.L.; Bauman, D.E.; Peel, C.J.; Steinhour, W.D. Effect of bovine somatotropin on metabolism of lactating dairy cows: Energy and nitrogen utilization as determined by respiration calorimetry. J. Nutr. 1988, 118, 1024–1030.

174. Fourichon, C.; Seegers, H.; Beaudeau, F.; Verfaille, L.; Bareille, N. Health-control costs in dairy farming systems in western France. Livest. Prod. Sci. 2001, 68, 141–156.

175. Lawson, L.G.; Agger, J.F.; Lund, M.; Coelli, T. Lameness, metabolic and digestive disorders, and technical efficiency in Danish dairy herds: A stochastic frontier production function approach. Livest. Prod. Sci. 2004, 91, 157–172.

176. Lean, I.J.; van Saun, R.; DeGaris, P.J. Energy and protein nutrition management of transition dairy cows: Metabolic diseases of dairy cattle. Vet. Clin. North Am. Food Anim. Pract. 2013, 29, 337–366.

177. Von Keyserlingk, M.; Martin, N.P.; Kebreab, E.; Knowlton, K.F.; Grant, R.J.; Stephenson, M.; Sniffen, C.J.; Harner, J.P.; Wright, A.D.; Smith, S.I. Invited review: Sustainability of the US dairy industry. J. Dairy Sci. 2013, 96, 5405–5425.

178. Sundrum, A. "Healthy food" from healthy cows. In Organic Farming and Food Production; Konvalina, P., Ed.; InTech: Rijeka, Croatia, 2012; pp. 95–120.

179. Kahneman, D. Thinking, Fast and Slow; Macmillan: New York, NY, USA, 2011.

180. Deutsche Forschungsgemeinschaft. Future Perspectives of Agricultural Science and Research: Perspektiven der Agrarwissenschaftlichen Forschung; Wiley-VCH: Bonn, Germany, 2005.

181. McNamara, J.P. Ruminant Nutrition Symposium: A systems approach to integrating genetics, nutrition, and metabolic efficiency in dairy cattle. J. Anim. Sci. 2012, 90, 1846–1854.

182. Friggens, N.C.; Newbold, J.R. Towards a biological basis for predicting nutrient partitioning: The dairy cow as an example. Animal 2007, 1, 87–97.

183. Khan, M.J.; Hosseini, A.; Burrell, S.; Rocco, S.M.; McNamara, J.P.; Loor, J.J. Change in subcutaneous adipose tissue metabolism and gene network expression during the transition period in dairy cows, including differences due to sire genetic merit. J. Dairy Sci. 2013, 96, 2171–2182.

184. Ideker, T.; Galitski, T.; Hood, L. A new approach to decoding life: Systems biology. Annu. Rev. Genomics Hum. Genet.2001, 2, 343–372.

185. Loor, J.J.; Bionaz, M.; Drackley, J.K. Systems physiology in dairy cattle: Nutritional genomics and beyond. Annu. Rev. Anim. Biosci. 2013, 1, 365–392.

186. Van Dien, S.; Schilling, C.H. Bringing metabolomics data into the forefront of systems biology. Mol. Syst. Biol. 2006, 2.

187. Bannink, A.; France, J.; Dijkstra, J. Modeling approaches to link various levels of organization in animal physiology. In Systems Biology and Livestock Science; Te Pas, M.F.W., Bannink, A., Woelders, H., Eds.; Wiley-Blackwell: Ames, IA, USA, 2011; pp. 191–214.

188. Loor, J.J.; Bionaz, M.; Invermizzi, G. Systems biology and animal nutrition: Insights from the dairy cow during growth and the lactation cycle. In Systems Biology and Livestock Science; Te Pas, M.F.W., Bannink, A., Woelders, H., Eds.; Wiley-Blackwell: Ames, IA, USA, 2011; pp. 215–246.

189. Hampf, M.; Gossen, M. Promoter crosstalk effects on gene expression. J. Mol. Biol. 2007, 365, 911–920.

190. Kaern, M.; Elston, T.C.; Blake, W.J.; Collins, J.J. Stochasticity in gene expression: From theories to phenotypes. Nat. Rev. Genet. 2005, 6, 451–464.

191. Mossio, M.; Bich, L.; Moreno, A. Emergence, closure and inter-level causation in biological systems. Erkenntnis 2013,78, 153–178.

CHAPTER 5

Control of Mammary Function During Lactation in Crossbred Dairy Cattle in the Tropics

Narongsak Chaiyabutr[1]

[1] Department of Physiology, Faculty of Veterinary Science, Chulalongkorn University, Bangkok, Thailand

1. INTRODUCTION

The major problems for the dairy practices in the tropical country are low milk yield and short lactation period of either pure exotic or crossbred dairy cattle. Many factors affect milk production in dairy cattle in tropical areas including high environmental temperature and humidity, lower genetic potential for milk production in indigenous cattle and inadequate supply of food during the dry and hot summer season. Crossbreeding of indigenous and exotic cattle for tropical use has been exploited as an efficient tool for blending the adaptability of native cattle with the high milking potential of exotic breeds resulting in increased milk production. In crossbred cattle, mechanisms of milk secretion are known to be inherited and are thought to be among the causes of differences in metabolic parameters. Milk secretion is a continuous process and requires a continuous supply of substrates for milk productions. Short persistency of lactation is the low ability of cow to continue producing milk at a high level after the peak of lactation. Improved persistency of lactation can contribute to decreasing the cost of the production system because lactation persistency is associated with feeding and health costs including reproductive performance. (Sölkner &

Fuchs,1987). There are many different ways to improve persistency of lactation, e.g. during summer, different types of crop residues have been used to feed animals as roughage in attempting to improve dairy productivity during period of scarcity (Jayasuriya & Perera, 1982; Promma et al., 1994). However, there is considerable evidence that understanding the mechanism acting within the body in relation to the control of milk secretion may improve of milk production in crossbred dairy cattle in tropics.

It is well known that mammary growth during pregnancy is a prerequisite for satisfactory lactation in all mammals. During pregnancy, the mammary gland is competing with many other organs for nutrition to sustain growth. Mammary secretory activity in dairy cattle is initiated from pregnancy although it is at a low level. Experimental studies in both periods of pregnancy and lactation will give some clue as to the nature of the developed mechanism of the mammary gland. During pregnancy, maternal bodily functions are altered, e.g. cardiac output and heart rate, while during lactation, many bodily functions are also altered, e.g. general circulation and body fluid (Hanwell & Peaker, 1977). Therefore, knowledge of extra-mammary factors influencing milk production is essential to study in both pregnancy and period of lactation which can provide quantitative information in crossbred Holstein cattle

The lactating mammary gland is dependent upon its blood supply to provide substrates at appropriate rates to sustain milk synthesis. The rate of substrates supplying to the mammary gland is determined by substrate concentration in the plasma and mammary blood flow. There is evidence that substrate supply to the mammary gland is often inadequate to maintain the maximum rate of milk synthesis. This raises the question: do changes in bodily function (water balance, general circulation and mammary circulation) alone have an effect on milk secretion or is the effect solely due to inadequate of the utilization of substrates in the mammary gland in crossbred dairy cattle? As glucose is the principal precursor of lactose, the decrease in milk lactose can be explained by a change of the mammary utilization of glucose (Faulkner & Peaker, 1987). Lactose is a highly osmotic component, which allows the drainage of water from blood to the alveolar compartment. As such, it is the principal milk component regulating the volume of milk production. Glucose is known to play an important role not only in lactose synthesis but also providing the reducing equivalent required for the synthesis of fatty acid *de novo* in the mammary gland (Chaiyabutr et al., 1980). Very few data are available regarding the dynamics and the regulation of glucose metabolism in whole body and the mammary gland of different types of crossbred cattle. Insight into the study in the utilization of glucose metabolism for synthesis of milk components in different metabolic pathways in the mammary gland, have improved understanding of what factors influencing the effect of low milk yield of crossbred dairy cattle in the tropics.

The role of endocrine regulation in initiation and maintenance of lactation is known to occur in many species. However, hormonal requirement among mammalian species differ considerably, for example, in rabbit prolactin alone can maintain lactation, while in cows prolactin is not a rate limiting hormone in established lactation in place growth hormone become relatively more important (Hart, 1973;Mepham, 1993). Very few data are available in the study of circulating hormones during lactation in crossbred dairy cattle. The circulating concentrations of some hormones will be expected to change and relate to the mechanism responsible for the control of milk secretion in different types of crossbred Holstein cattle. Investigation of the plasma levels of various hormones (thyroid hormone, prolactin, cortisol, growth hormone, insulin, glucagon, progesterone and estradiol) involving milk secretion in late pregnancy and different stages of lactation may give some information on the control mechanism of milk secretion either directly or indirectly on the function of mammary gland in different types of crossbred Holstein cattle.

Thus, this chapter is intended to review the highlight the regulatory mechanisms underlying marked low milk yields, by providing an updated summary of the results obtained concerning the physiological changes in both extra-mammary factors and intra-mammary factors of crossbred Holstein cows between 50HF:50Red Sindhi(50%HF) and 87.5HF:12.5RS(87.5%HF) cattle in the tropics

2. WATER METABOLISM AND MAMMARY CIRCULATION IN DIFFERENT STAGES OF LACTATION OF CROSSBRED CATTLE

It is known that lactating dairy cows metabolize large amounts of water and are affected rapidly by water deprivation (Murphy 1992). In the tropics, dairy cattle utilize body water to maintain homeostasis in the mechanisms of thermoregulation and lactation. The effect of excessive heat load in high environment temperature leading to increase in heat dissipation (Hahn et al., 1999), which may lead to a loss body water and decrease in milk yield in lactating cattle. However, an increase in water intake during lactation closely match to increase in water secrete in milk, since milk composition has about 87% of water. During lactation, many bodily functions are altered; for example, blood volume and cardiac output are increased (Hanwell & Peaker 1977). These changes may effectively alter body fluid and thus circulatory distribution including the blood supply to the mammary gland.

2.1. Water Metabolism

2.1.1. Measurement of Water Metabolism

In this section, a brief description of methods for measuring water metabolism is presented. According to the study of Chaiyabutr et al., (1997) in measurements of water metabolism, the water turnover rate (WTO), total body water (TBW) in dairy cattle were performed using tritiated water dilution techniques. Briefly, the dairy cattle was injected intravenously with carrier free tritiated water in normal saline at a single dose of 3,000 μCi per animal. The equilibration time was determined by taking serial venous blood samples for 3 days after the injection. Blood samples were collected at 20, 30, 40, 50, 60 min and 4, 8, 20, 26, 32, 44, 50, 56, 68 and 74 h subsequent to the injection. Preparation of samples for counting was achieved by the internal standardization technique. The corrected activity of samples, in disintegrate per minute (d.p.m.), were plotted on semi-logarithmic paper against time. The dilution curve of tritiated water in plasma was described by an exponential equation using a one compartment model, which is

$$Yi = Ae\text{-}^{klt},$$

where Y is concentration of tritium in plasma at time t (nci/ml); A is plasma concentration intercept 1 in nci/ml.

The extrapolated activity at theoretical zero time of complete mixing of radio-isotope was used to determine the total body water space (TOH). The TOH space was calculated:

TOH space (ml) = [standard count (dis/min) x dose (ml)] / [radio activity counts at zero time (dis/min)].

The biological half-life of tritium-labelled water (T 1/2) was determined from the slope of the linear regression line obtained from plot on semi-logarithmic paper of the activity of the samples taken over the period of 3 days against time. The water turnover rate was calculated from the equation:

WTO (l/day) = 0.693 x TOH space / T 1/2.

Total body water (TBW) was calculated by using the corrected factor (1 - fraction of plasma solids) x TOH space (Chaiyabutr et al. 1997).

Work on water metabolism has produced some interesting results. The regulation of body fluids and mammary circulation in different types of crossbred Holstein Friesians (HF) cattle showed some differences between 50%HF and 87.5%HF cattle during late pregnancy and lactation (Chaiyabutr et al., 1997). A short persistency of milk yield during the transition period from early to mid lactation has been observed in 87.5%HF cattle. In contrast to 50%HF cattle, persistent lactation seemed to be apparent throughout periods of lactation (Chaiyabutr et al. 2000a). The different control mechanisms were at play in the regulation of milk production as lactation advances in 50%HF and 87.5%HF cattle. The 87.5%HF cattle had lower efficiency in water retention mechanism and poor adaptation to tropical environment in comparison to 50%HF (Chaiyabutr et al., 1997; 2000a). The high genetic similarity of 87.5%HF cattle to the exotic *bos taurus* breed may lead to poor adjustment to the tropical environment, while yielding high milk production in early lactation. However, the milk yield of 87.5%HF cattle has been shown to be higher in early lactation than 50%HF cattle while the ratio of total dry matter intake to milk yield was lower in 87.5%HF as compared to 50%HF cattle. The energy output in milk and for maintenance for 87.5%HF cattle was greater than the energy consumed in the food in early lactation. The 50%HF cattle was approximately in energy equilibrium, there being no change in the ratio of dry matter intake to milk yield among periods of lactation.

Water is known to play a prominent role in the processes involved in the formation of milk, since milk is isoosmotic with plasma. In intra-mammary process, the net transfer of water from plasma to the alveolar lumen is directly proportional to the transfer of solute. Water is drawn into secretory vesicles because of the presence of osmotically active lactose (Kaufmann & Hagemeister,1987; Linzell & Peaker, 1971),which is formed in the golgi apparatus of the lactating cells. Thus, the amount of water secreted into milk is regulated by the rate of lactose synthesis. However, cellular water metabolism in lactating cell is not a final story of the processes of milk formation. The extra-mammary factor about body water content will be a prerequisite for satisfactory lactation, which may influence on lactation persistency in crossbred cattle.

Measurements of water turnover rate and total body water using tritiated water as a marker in different stages of lactation of crossbred Holsteins cattle is shown in Fig.1 (Chaiyabutr et al.,1997; 2000a).Chaiyabutr et al (2000a) have noted that the total body water as a percentage of body weight of 87.5%HF cattle was lower than 50%HF cattle in all periods of lactation. This may be attributed to a relatively lower efficiency in the water retention mechanism, although the water intake was higher in 87.5%HF animals during high milk yield. Low water content may be related to the poor adaptation of 87.5%HF to the tropical environment. Poorer lactation persistency in higher yielding cows has also been noted (Chase, 1993; Coulon et al., 1995). The studies in both 87.5%HF and 50%HF housing in the same shed under the same environment by Chaiyabutr et al., (2000a) have shown that a lower water turnover rate was apparent especially in late lactation of the 87.5%HF cattle in comparison with 50%HF cattle.

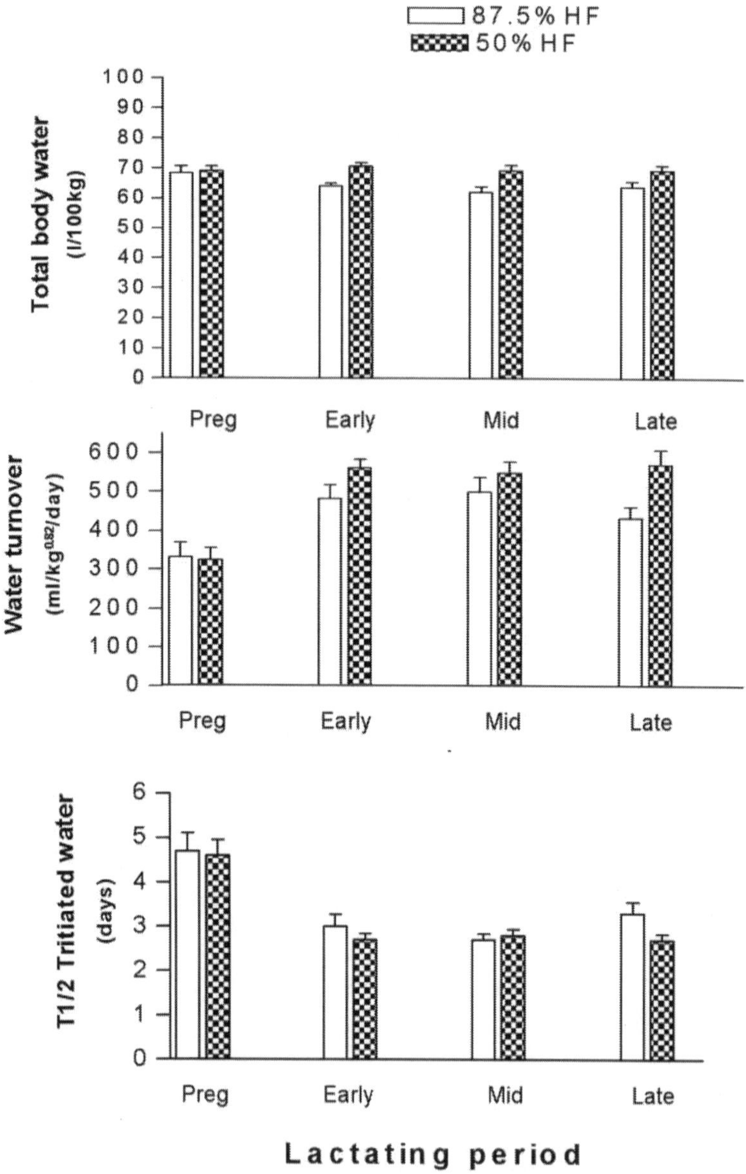

Figure 1. Water turnover rate, total body water and half-life of tritiated of water during late pregnancy and lactation of 87.5%HF and 50%HFcattle in tropical environment. (Data from Chaiyabutr et al., 1997;2000a).

The water turnover rate of both types of crossbred cattle was not influenced by environmental conditions, although marked differences of water turnover rate and half-life of body water in animals have been reported during the winter and summer (Ranjhan et al., 1982). A higher water reserve in 50%HF cattle would not only provide a higher reservoir of soluble metabolites for biosynthesis of milk but was also useful in slowing down the elevation in body temperature of this breed during lactation in hot conditions (Nakamura et al.,1993). The 87.5%HF cattle did not require greater amounts of water, while it could restore their body fluids to equilibrium in all lactating periods. The differences between crossbred and purebred animals in body composition and water turnover rate have also been reported (Macfarlane & Howard 1970). The study of the regulation of body fluid and mammary circulation during late pregnancy and early lactation of crossbred Holstein cattle byChaiyabutr et al., (1997) have found that the water turnover rate was significantly higher approximately 45 % in lactating cattle than in pregnant cattle in either 87.5%HF or 50%HF cattle. During the lactating period the half-life of tritiated water (2.9 days) was significantly lower than pregnant period (4.7 days).

2.2. Mammary Circulation and Blood Volume

2.2.1. Measurement of Udder Blood Flow

According to the study of Chaiyabutr et al., (1997), the measurement of udder blood flow in dairy cattle was described. Briefly, blood flow through half of the udder was determined by measuring the dilution of dye T-1824 (Evans blue) by a short term continuous infusion in the milk vein of dairy cattle. A dye (T-1824) was dissolved in sterile normal saline and diluted to a concentration of 100 mg/l. The solution was infused by a peristaltic pump at a constant rate of 80 ml/min into the milk vein for 1-2 min. Before infusion, blood was drawn from downstream in the milk vein as a pre-infusion sample. About 10 seconds after starting infusion, 10 ml of blood was drawn from downstream in the milk vein at a constant rate into a heparinized tube. Two consecutive plasma samples were taken during dye infusion. Blood flow of half of the udder was calculated from the concentration of dye in plasma samples using the equation derived by Thompson & Thomson (1977). In lactating cows, quarter milking showed that the yields of the two halves of the udder were similar. Udder blood flow was therefore calculated by doubling the flow measured in one milk vein.

2.2.2. Measurement of Plasma Volume and Blood Volume

In the studies of plasma volume and blood volume, plasma volume was measured by dilution of Evan's blue (T-1824) dye. The injection of 20 ml of the dye (0.5 g/100 ml normal saline) into the ear vein catheter was performed and venous blood samples were collected from the jugular vein at 30, 40 and 50 min after dye injection. Dilution of dye at zero time was determined by extrapolation. Blood volume was calculated from the plasma volume and packed cell volume (PCV). Plasma osmolality was measured using the freezing point depression method.

The results of study by Chaiyabutr et al (2000a)have noted that the mammary circulation of crossbred dairy cattle between 87.5%HF and 50%HF varied during different periods of lactation. Mammary blood flow is known to be a major determinant controlling milk production in a way to carry milk precursors to the mammary gland at the process of milk synthesis. A decrease in blood flow to the mammary gland coinciding with a short persistency of milk yield during the transition period from early to mid lactation occurred in 87.5%HF cattle (Chaiyabutr et al. 2000a). The milk yield of 87.5%HF cattle significantly declined in mid-lactation from early lactation. In contrast to 50%HF cattle, short persistent lactation seemed to be apparent throughout periods of lactation (Fig. 3). The study for alteration of lactation persistency in different crossbred animals will throw some light on a useful index for studying adaptability in crossbred cattle, which will provide information on choosing suitable crossbred dairy cattle in the tropics. A higher mammary blood flow during early lactation compared to mid-lactation in 87.5%HF cattle could not be attributed to a change in blood volume and plasma volume, which remained nearly constant (Fig.2). Plasma osmolality remained unchanged during the course of lactation in both 87.5%HF and 50%HF cattle indicating that homeostasis was being maintained throughout all periods of lactation. Such differences of the mammary circulation between 87.5%HF and 50%HF animals can be attributable to disparities in breed. The ratio of mammary blood flow to the rate of milk yield did not change during the course of lactation in 50%HF animals. The marked decrease in the mammary blood flow of 87.5%HF at mid-lactation would correlate with the decrease of milk yield. The decline in milk yield after the peak will be due primarily to a decreased availability of substrates to the mammary gland. The marked increase in the ratio of mammary blood flow to the rate of milk yield during lactation advance to mid-lactation (548:1) and late lactation (668:1) in 87.5%HF cattle indicated a more decreased secretory activity of mammary tissue. The question then arises as to whether mammary metabolism influences mammary blood flow or mammary blood flow influences mammary metabolism.

Many studies on mechanisms concerned with regulation of mammary blood flow, local and extra-mammary production of vasoactive agents and activity of mammary sympathetic nerves have been reviewed comprehensively (Linzell, 1974). The study on mammary circulation in relation to the general circulation of crossbred cattle at different stages of lactation is shown in Fig. 3. The mean values of mammary resistance (the ratio of mean arterial pressure to mammary blood flow) did not significant change in different periods of lactation in either 87.5%HF or 50%HF cattle. It is clear that the local changes for vasoconstriction in the udder were not apparent in different periods of lactation for both types of crossbred cattle. This proposal was indicated that humoral factors were responsible, especially the level of plasma growth hormone in the regulation of mammary circulation and milk production between 87.5%HF and 50%HF cattle. A number of studies have demonstrated that similar proportion increases in milk secretion and mammary blood flow occurred during growth hormone treatment in goats and cows (Hart et al., 1980; Chaiyabutr et al 2007). A marked decline in blood flow to the mammary gland during the transitional period from early lactation to mid-lactation in 87.5%HF cattle with the constancy of the plasma volume and blood volume were apparent (Chaiyabutr et al 2000a). Regulation of mammary blood flow from the early lactation to mid-lactation in 87.5%HF cattle may be regarded a major homeorhetic principle (Bauman & Currie, 1980).

Mammary growth during pregnancy has been known to be a prerequisite for satisfactory lactation. The degree of local vasoconstriction was greater during pregnancy and less in lactation for both types of crossbred cattle. In late pregnancy, the mammary circulation of 50%HF was less than 87.5%HF cattle which could be due to variations in the developments of mammary blood vessels and mammary cells. The high genetic blood level closing to the exotic *bos taurus* breed of 87.5%HF animals may cause a rapid rate in the development of the secretory activity of the mammary cell in the late pregnancy. Local vasodilatation produced by the active cells (Hanwell & Peaker, 1977; Lacasse et al., 1996) would decrease in the resistance of the vascular bed and a higher mammary blood flow of 87.5%HF in comparison with 50%HF cattle (Chaiyabutr et al., 2000a). These physiological changes would accompany with increases in plasma volume and blood volume in the lactating period when compared with the pregnant period in either 87.5%HF or 50%HF cattle. These changes are agree to other reports that blood volume, plasma volume and water turnover are markedly higher in lactating animals than in pregnant animals (Hanwell & Peaker, 1977; MacFarlane & Howard, 1970). The packed cell volume significantly decreased in the lactating period of 87.5%HF animals while it did not significantly change in 50%HF animals. The packed cell volume of 50%HF animals was significantly higher than that of 87.5%HF animals.

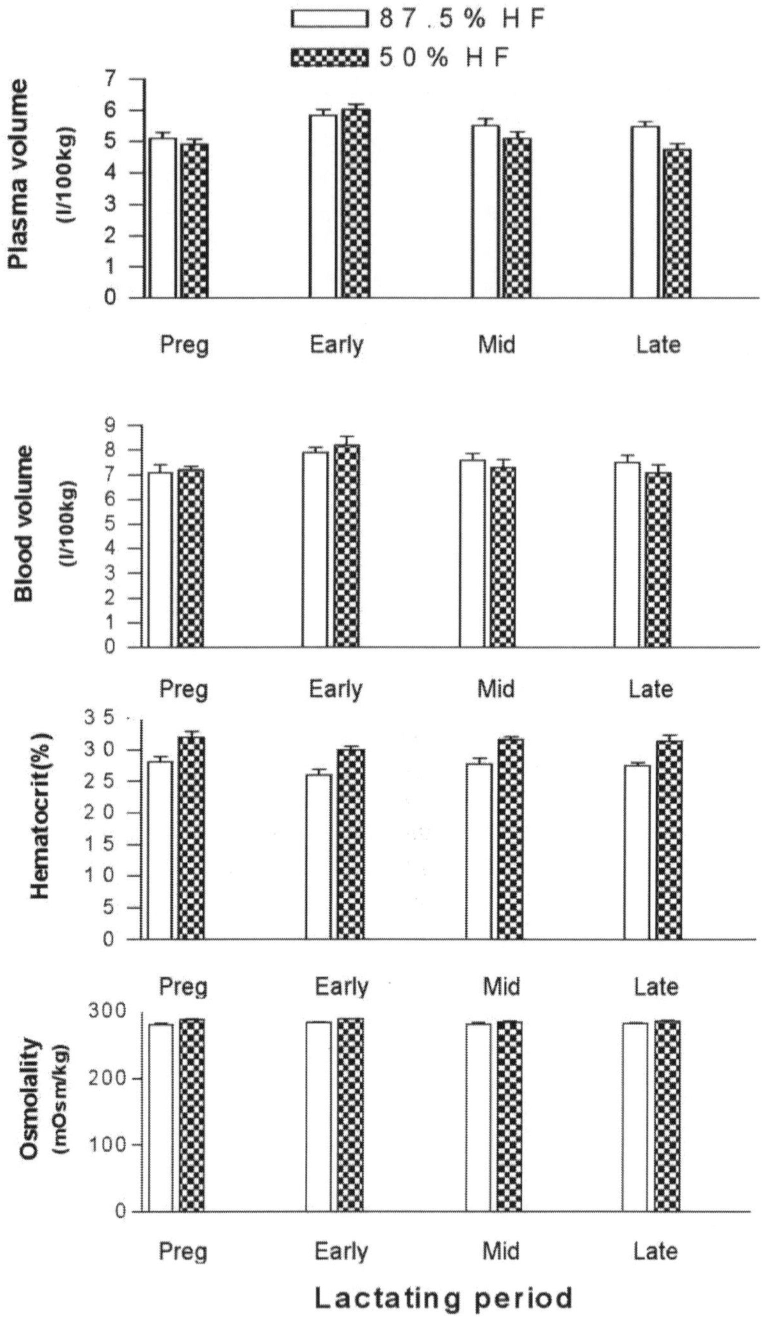

Figure 2. Plasma volume, blood volume, hematocrit and plasma osmolality during late pregnancy and lactation of 87.5%HF and 50%HF cattle in tropical environment. (Data from Chaiyabutr et al., 1997; 2000a).

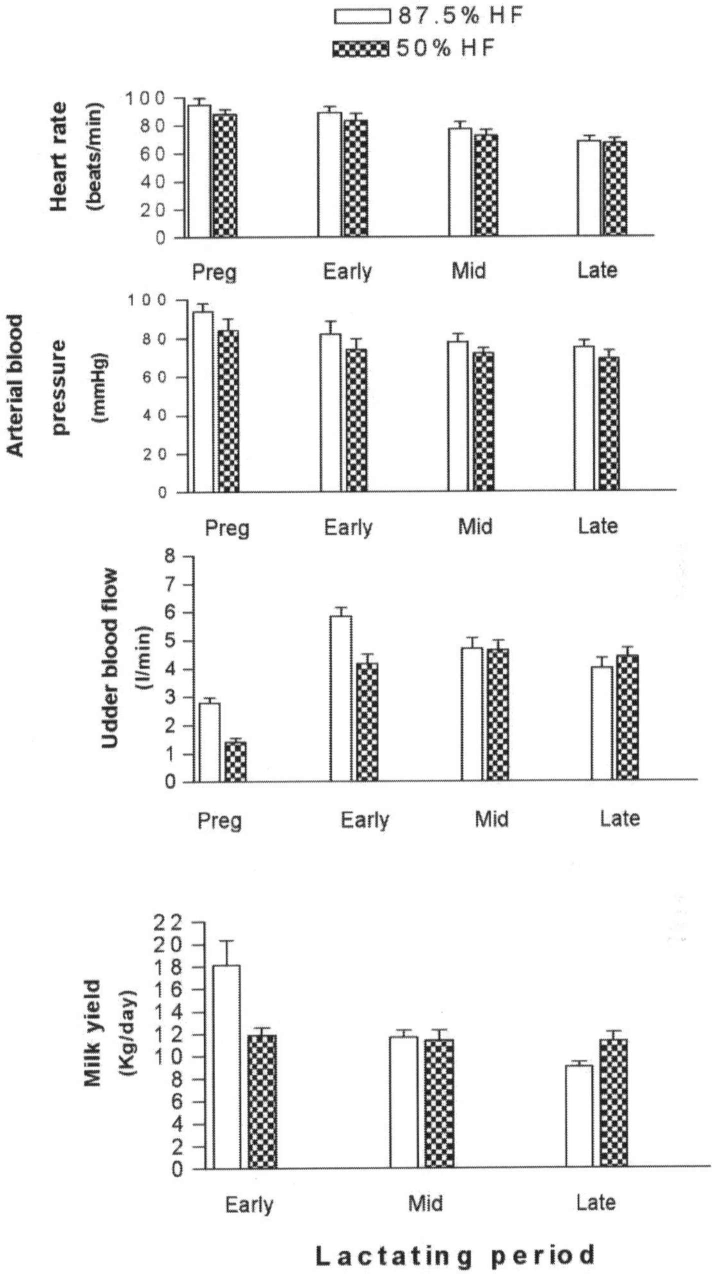

Figure 3. The mean arterial pressure, heart rate, udder blood flow and milk yield during late pregnancy and lactation of 87.5%HF and 50%HF cattle in tropical environment. (Data from Chaiyabutr et al., 1997; 2000a).

There were no significant differences in heart rate, arterial blood pressure and plasma osmolality between the periods of late pregnancy and early stage of lactation in either 87.5%HF or 50%HF cattle. Udder blood flow was nearly three times higher in lactating period than in late pregnant period in either 87.5%HF or 50%HF cattle. The udder blood flow of 87.5%HF was significantly higher while mammary resistance was significantly lower than in 50%HF animals. Milk secretion in early lactation of 87.5%HF animals was higher than that of 50%HF animals. The ratio of DM intake to milk production for 87.5%HF animals was lower than that of 50%HF animals. From these results it can be concluded that the difference between breeds was found in 87.5%HF animals which had a higher milk yield but a lower adjustment for the regulation of body fluids during pregnancy and lactation in comparison to 50%HF animals.

3. ROLE OF SUBSTRATES SUPPLY FOR MILK PRODUCTION

It is known that the rate of substrates supplying to mammary gland is determined by substrate concentration in the plasma and mammary blood flow. The sustainable milk synthesis in the mammary gland is dependent upon its blood supply to provide substrates at appropriate rates. There is evident that substrate supply to the mammary gland is often inadequate to maintain the maximum rate of milk synthesis (Linzell & Mepham 1974). The mammary gland may be producing milk at a rate below its potential. The rate of milk production depends on function of number of secretory cells and their metabolic activity. However, the mechanisms can be devided into three main levels of regulation, which are the arterial flow of substrates in the mammary gland, the substrate extraction by the mammary gland and the metabolic and secretion activities of the mammary epithelial cell.

3.1. Changes of Substrates Supply

3.1.1. Measurement of Mammary Uptake of Substrates

According to the study of Chaiyabutr et al., (2002), the measurement of udder blood flow in dairy cattle was described. Briefly, The uptake of substrate by the udder (U_S), expressed as μmol/min, was calculated from the equation:

$$U_S = MPF \times (P_A - P_V)$$

The substrate extraction by the mammary (E_S):

$$E_S = (P_A - P_V) / P_A$$

Where:

MPF = Mammary plasma flow (ml/min)

P_A = Concentration of substrate in coccygeal arterial plasma (µmol/ml)

P_V = Concentration of substrate of plasma from milk vein (µmol/ml)

The results of study by Chaiyabutr et al (2002) in both 87.5%HF and 50%HF cattle have shown that the mammary uptake of many substrates between late pregnancy and lactating periods in crossbred HF cattle was not based on changes in arteriovenous concentration differences and extraction ratio. Mammary growth during pregnancy has been known to be a prerequisite for satisfactory lactation. The pattern of mammary growth varied during pregnancy and early lactation for both 87.5%HF and 50%HF cattle. Changes of substrates across the udder will account for changes in mammary blood flow. Both 87.5%HF and 50%HF cattle showed the low values of both arteriovenous concentration differences of substrates across the mammary gland and mammary extraction ratio during late pregnancy, which differ from the lactating period with a marked increase of mammary uptake of substrates for milk synthesis.

Volatile fatty acid in form of acetate is known to be the major of source energy of normal fed ruminants. No differences in mammary arteriovenous concentration differences and extraction ratio of acetate have been shown between late pregnancy and lactation in both 87.5%HF and 50%HF cattle. The low rate of mammary uptakes for acetate during pregnancy was dependent upon rate of blood flow. During late pregnancy, circulating β-hydroxybutyrate level which arise mainly from rumen butyrate in the normal fed animal (Leng & West, 1969). There were no obvious of arteriovenous concentration differences and extraction ratio of β-hydroxybutyrate across the mammary gland of both 50%HF and 87.5%HFcattle.

The concentrations of arterial plasma free fatty acid and triacylglycerol increased in the late pregnancy of both 50%HF and 87.5%HFcattle and were more sensitive to alteration than other blood substrates; this phenomenon has been proposed as an indication of under-nutrition (Reid & Hinks, 1962). The mobilization of body fat in late pregnancy occurs in response to hormonal secretion (Lindsay, 1973). It would be a physiological phenomenon, not a consequence of under-nutrition. The measurement of arteriovenous differences of FFA across the mammary gland together with mammary blood flow did not provide a quantitative estimation of their total uptake by mammary tissue in either 50%HF and 87.5%HF cattle, since the release of FFA into venous blood due to triacylglycerol hydrolysis during the uptake of plasma triacylglycerol as in lactation has been noted (West et al., 1967). During the development of the mammary gland in normal late pregnancy, an occurring the utilization of FFA may cause changes in arteriovenous differences, the extraction ratio and net uptake of FFA. The net uptake of triacylglycerol by the mammary gland significantly increased in lactating period in comparison to the late pregnancy in both 87.5%HF and 50%HF cattle. It is possible that a change for releasing of FFA is a result of changes of enzymatic activity of lipoprotein lipase in the mammary tissue. This enzyme activity has been reported to be higher in lactating bovine mammary tissue relative to pregnant tissue (Shirley et al., 1973).

3.2. Glucose Metabolism And Glucose Utilization In The Mammary Gland

It is known that an increase of milk yield can be achieved only by increasing the rate of lactose synthesis. The lactating udder utilizes most of the glucose entering the circulation of ruminants and irreversible glucose loss from plasma is highly correlated with lactose output (Bickerstaffe et al., 1974;Horsfield et al., 1974). The role of glucose in regulating milk secretion has also been demonstrated in the isolated perfused udder (Hardwick et al., 1961). The regulation of the milk yield of animals is therefore mainly based on the mechanisms governing the quantity of glucose extracted by the udder and converted into lactose. The metabolic fate of glucose metabolism will consider the utilization in the whole body relating to the utilization in the mammary gland in either pregnancy or lactation.

3.2.1. Measurements of Glucose Kinetic And Glucose Utilization In The Mammary Gland

According to Chaiyabutr et al.,(2000c), body glucose metabolism and intramammary glucose metabolism has been described. In brief, glucose kinetic studies at different stages of lactation were carried out by using continuous infusion of both [U-^{14}C]-glucose and [3-^{3}H]-glucose solution. A priming dose of radioactive glucose in 20 ml of normal saline containing 30 μCi [3-^{3}H]-glucose and 15 μCi[U-^{14}C]-glucose was administered intravenously via the ear vein catheter and followed by a continuous infusion(using Peristatic pump) of 1 ml/min of normal saline solution (0.9%) containing 0.7 μCi/ ml of [U-^{14}C]-glucose and 1.5 μCi/ml of [3-^{3}H]-glucose for 3 h. During the last 1 h of continuous infusion, three sets of blood samples were collected at 20 min intervals.

The glucose turnover rate in the whole animal (T), expressed as μmol/min, was calculated from the equation:

$$T = I/G_A$$

where, I = rate of infusion of [U-^{14}C]glucose or [3-^{3}H]glucose (μCi/min) and G_A = specific activity of ^{14}C- or ^{3}H-glucose in arterial plasma at equilibrium (μCi/μmol).

Recycling of glucose carbon in the whole animal, expressed as % glucose turnover, was calculated from the equation:

$$\text{Recycling} = (T_3 - T_{14}) \times 100 / T_3$$

Where:

T_3 = Reversible turnover of glucose calculated from [3-^{3}H]glucose

T_{14} = Irreversible turnover of glucose calculated from [U-^{14}C]glucose

Uptake of glucose by the udder (U_G), expressed as μmol/min, was calculated from the equation:

$$U_G = MPF \times (P_A P_V)$$

Where:

MPF = Mammary plasma flow (ml/min)

P_A = Concentration of glucose in coccygeal arterial plasma (μmol/ml)

P_V = Concentration of glucose of plasma from milk vein(μmol/ml)

The milk component output (MO), expressed as µmol/min and was calculated from the equation:

$$MO = Ms \times Cc/1000$$

Where:

Ms = Milk secretion rate (ml/min)

Cc = Concentration of components in milk (µmol/ L)

Incorporation (A) of radioactivity from glucose into milk components was calculated from the equation:

$$A = M_A/G_A \times t$$

Where:

A = Incorporation of radioactivity from glucose into milk components (µmol/min)

M_A = Total activity of ^3H or ^{14}C in the milk components (µCi)

G_A = Specific activity of ^{14}C-or ^3H-glucose in arterial plasma at equilibrium (µCi/µmol)

t = Time of infusion (min)

On the glucose kinetics studies using 3-[^3H] glucose and [U-^{14}C] glucose infusion, the total glucose entry rate (the reversible turnover rate of 3-[^3H] glucose) and utilization rates of glucose (the irreversible turnover rate of [U-^{14}C] glucose) increased significantly during early lactation in comparison to late pregnancy for both 50%HF and 87.5%HF cattle (Fig.4). Recycling of glucose-C was approximately 20% in both crossbred cattle which was unaffected by the stage of late pregnancy or early lactation. During late lactation, glucose carbon recycling and plasma glucose clearance increased in both 50%HF and 87.5%HF cattle. Comparing 50%HF and 87.5%HF cattle, arterial plasma glucose concentrations were slightly higher during pregnant period but significantly higher in lactating periods in 50%HF cattle (Chaiyabutr et al.,1998). This indicates that steady state conditions between the rate of utilization of glucose and the rate of gluconeogenesis differ in the body pool of glucose. The uptake, arteriovenous differences and extraction ratio for glucose across the udder increased in the lactating period in either 87.5%HF or 50%HF cattle. Glucose uptake by the udder accounted for 65% of the total glucose turnover rate in 87.5%HF cattle and 46% in lactating 50%HF cattle, although both crossbred cattle exhibited the same body glucose turnover rate (Chaiyabutr et al., 2000b). The 87.5%HF cattle had high milk yield, body glucose metabolism and udder glucose metabolisms as compared with 50%HF cattle in early lactation.

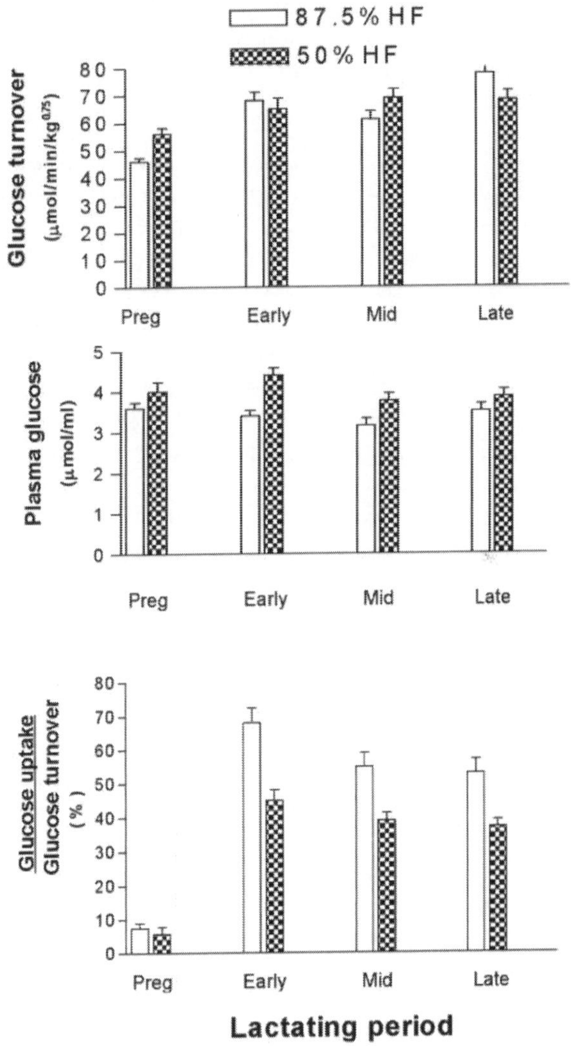

Figure 4. Glucose turnover, plasma glucose concentration and percentage of glucose uptake to glucose turnover during late pregnancy and lactation of 87.5%HF and 50%HF cattle in tropical environment. (Data from Chaiyabutr et al., 1998; 2000b).

In early lactation, the high milk yields and lactose secretion of 87.5%HF cattle were related to glucose uptake by the udder and udder blood flow as compared with 50%HF cattle. An increase in udder blood flow during early lactation was higher in 87.5%HF cattle than in 50%HF cattle. The marked decreases in udder blood flow, glucose uptake, lactose secretion and milk yield of 87.5%HF cattle were apparent in mid- and late lactation. In contrast to 50%HF cattle, no alterations for udder blood flow, udder glucose uptake, lactose secretion and milk yields were apparent throughout the course of lactation. The 87.5%HF cattle had a higher milk yield but a shorter peak yield and poorer persistency in comparison with 50%HF cattle (Chaiyabutr et al., 2000a). Changes in the utilization of glucose by the mammary gland for milk production in crossbred cattle would be dependent on changes in intramammary factors.

In general the rate of milk secretion is thought to be determined primarily by the secretion of lactose and consequent osmotic water movements follows lactose which create milk secretion as whole (Rook & Wheelock, 1967; Linzell & Peaker, 1971). An increase in milk yield can be attributed to an increase in the rate of lactose synthesis. Glucose is essential for milk secretion (Hardwick, Linzell & Price, 1961) and glucose moiety of lactose arises directly from plasma glucose (Ebner & Schanbacher, 1974). No differences of mammary arteriovenous differences of glucose in different periods of lactation in either 87.5%HF or 50%HF cattle (Chaiyabutr et al., 2000b, 2002) suggested that glucose uptake was determined mainly by mammary blood flow(Linzell, 1973). Mammary blood flow, milk yield and lactose yield were not affected by alterations of plasma glucose concentrations in both 87.5%HF and 50%HF cattle (Chaiyabutr et al. 2000b). There is evidence to confirm that the plasma glucose concentration is not a major factor in determining lactose output. The major change would occur in the metabolism of the mammary epithelial cell in different periods of lactation. It might be reasonable to believe that the reduction of glucose uptake by the mammary gland during mid- and late lactation of 87.5%HF cattle would not be a result of the decrease in the rate of phosphorylation of glucose by hexokinase. The decrease in mammary glucose uptake has been shown to be insulin independent. The plasma insulin concentrations of both 87.5%HF and 50%HF cattle showed similar ranges in different periods of lactation (Chaiyabutr et al., 2000d). Transport of glucose in the mammary gland is known to be a rate-limiting step for milk synthesis which relates to glucose transporters in the acinar cell (Burnol et al., 1990). Thus, the extraction efficiency of glucose by the mammary gland would depend mainly on the capacity for transmembrane transport and/or intracellular metabolism of glucose. However, in mammary epithelial cells of ruminants, the GLUT1 transporter protein predominating in the bovine mammary gland under normal physiological conditions (Komatsu et al, 2005; Zhao et al., 1996) is probably never saturated. The regulation of glucose extraction by bovine mammary epithelial cells may occur at another level than transmembrane transport (Xiao & Cant, 2003).

According to Chaiyabutr et al. (2000c), the studies of glucose utilization in the mammary cell using 3-[^{3}H] glucose and [U-^{14}C] glucose as markers were performed in 87.5%HF and 50%HF cattle. Then, incorporation of radioactive glucose from blood into milk components, the uptake of glucose including other substrates and milk components output were calculated and expressed as µmol/min. in the metabolic pathway. As schemes shown in Fig. 5 & 6, the quantitative utilization of the glucose taken up by the mammary gland for the synthesis of lactose, metabolize via the pentose phosphate pathway, Embden-Meyerhof pathway and the tricarboxylic acid cycle showed differences between 87.5%HF and 50%HF cattle. The calculated percentage of metabolism of glucose 6-phosphate pool to galactose moiety of lactose during early lactation was about 92% in 87.5%HF cattle and 80% in 50%HF cattle (Chaiyabutr et al 2000c). The percentages of these values were consistent with the results reported in normally fed goats and bos Taurus cows (Wood et al., 1965; Linzell, 1968). During lactation advance to late lactation, the percentages of these values decreased markedly in 87.5%HF cattle. The synthesis of lactose usually involves combination of glucose and UDP-galactose; the UDP-galactose originates from glucose-6-phosphate (Ebner & Schanbacher, 1974). The availability of cytosolic glucose 6-phosphate in the cell is usually sufficient to account for the cytosolic lactose synthesis in all periods of lactation. In mid- and late lactation of 87.5%HF cattle, a decrease in the process of lactose biosynthesis might result of it impaired synthesis. The mammary gland cannot synthesize its own glucose, since the activity of glucose 6-phosphatase is low or absent in the mammary tissue (Davis & Bauman, 1974), reducing the chance for reversibility of the glucose 6-phosphate to glucose reaction. It is possible that a decrease in the process of lactose biosynthesis may occur at the last step catalyzed by lactose synthetase (Jones, 1978), and at the same time there is an increase in cytosolic concentration of glucose 6-phosphate. The reduction of glucose uptake during lactation advance in 87.5%HF cattle might be due to the inhibition of the enzyme hexokinase by high cytosolic concentration of glucose 6-phosphate pool. However, lactose biosynthesis is a complex process (Kuhn et al., 1980), there is still a need for more information to elucidate the changes in enzymatic activity in this particular system in different stages of lactation and in different breeds of cattle.

During the reduction in the lactose synthesis in mid- and late lactation in 87.5%HF cattle, there were no significant changes in the rate of fatty acid synthesis *de novo* in the mammary gland (Chaiyabutr et al 2000c). In contrast to 87.5%HF cattle, there was a low in the flux of glucose 6-phosphate through the pentose phosphate pathway in 50%HF cattle (Fig. 6). The glucose taken up by the mammary gland were metabolized in the pentose phosphate pathway about 10-21% in 87.5%HFcattle and 5-7% in 50%HF cattle in early to mid lactation (Chaiyabutr et al 2000c). Glucose metabolism accounted for the NADPH production from pentose phosphate pathway required for fatty acid synthesis *de novo* in the udder about 21% in 87.5%HF cattle and 15% in 50%HF cattle. The differences in glucose and lipid metabolism between 87.5%HF and 50%HF cattle would be accounted for by the fact of the different breeds of animals with different rates of udder metabolism. The high rate of glucose metabolism accounted for the NADPH production in the pentose phosphate pathway during early and mid-lactation, which also associated with the high transfer of carbon atom of glucose to glycerol for the esterification in milk fat. The percentages of glycerol synthesized from glucose carbon (using U-^{14}C glucose as marker) in vivo, were lower than those reported in studies in vitro by Hardwick and co-worker (1963), in which 23% of triacylglycerol glycerol was synthesized from glucose in the perfused goat udder. The difference can be attributed to the fact that the glucose carbon taken up by the mammary gland in intact conscious animals can be accounted for synthesis of lactose, citrate, CO_2 and lost as venous-plasma lactate. Some glucose may also be converted into amino acids for milk protein synthesis (Linzell & Mepham, 1968). In 87.5%HF cattle, the utilization of glucose carbon by mammary epithelial cells for the synthesis of triacylglycerol(using U-^{14}C glucose infusion) was lower in early lactation and conversely, the values expressed as percentages are higher during mid- and late lactation (Chaiyabutr et al 2000c). An increase in the proportion of ^{14}C-glucose converted to triacylglycerol was consistent with a large proportion of glucose 6-phosphate which was metabolized via the Embden-Meyerhof pathway (EMP) in late lactation in 87.5%HF cattle (Fig.5).

It is known that acetate, β-hydroxybutyrate, long chain fatty acids of triacylglycrol and free fatty acid fractions of plasma are major precursor of milk fatty acids. The balance data for the utilization of both short chain and long chain fatty acid were performed by calculating their likely contribution to milk free fatty acid knowing its composition and subtracting these values from the measured uptake of the substrates (Fig.5 & 6). Acetate and β–hydroxybutyrate were grouped together because both acetate and β–hydroxybutyrate were contributed to the synthesis of milk fatty acids up to and including C_{16}(Annison et al., 1968; Palmquist et al., 1969). In normally fed cattle, acetate is the main substrates for milk fat synthesis and β–hydroxybutyrate also make a more significant contribution to oxidative metabolism in the udder. There is a negligible oxidation of free fatty acid by the udder in normally fed animals (Bickerstaffe et al., 1974). The insignificant net uptake of free fatty acids by the mammary gland has been shown to be due to the simultaneous uptake of free fatty acids and release into mammary venous blood of fatty acid derived from plasma triacylglycerol which are hydrolysed during their uptake by the mammary gland (West et al., 1972). During mid- and late lactation of 87.5%HF animals there are marked decreases in utilization of both acetate and β–

hydroxybutyrate which are due to the decrease in their supplies from the blood stream (Chaiyabutr et al., 2002).

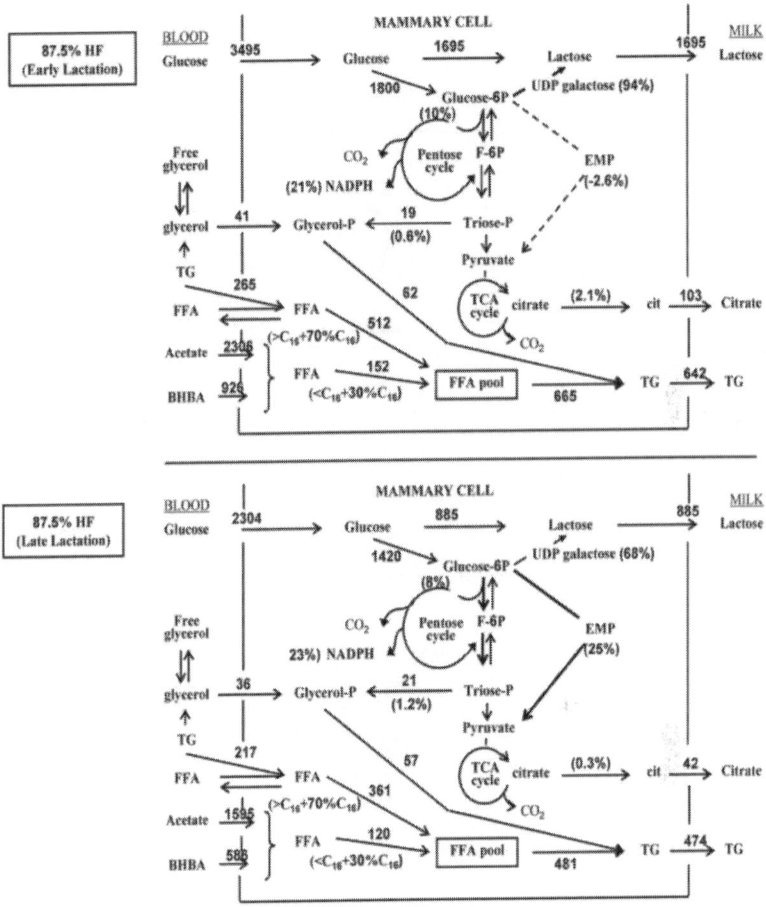

Figure 5. Scheme for calculated metabolic pathways involve intracellular partitioning of precursors of milk in mammary epithelial cells during early and late lactation of 87.5%HF cattle. Values are shown in micromole/min. (Calculated data from Chaiyabutr et al 2000c)

During early lactation, the milk fatty acid concentrations with a chain length of C_6 to C_{18} of 50% HF cattle were higher than those of 87.5% HF cattle feeding on similar diet (Chaiyabutr et al 2000c). During mid- and late lactation, similar milk fatty acid concentrations were maintained as in early lactation of 50% HF animals. There was considerable variation with advanced lactation in the level of milk fatty acid concentration of 87.5% HF cattle. During mid-lactation and late lactation, the milk fatty acid concentration, particularly with a chain length of C_{16} to C_{18}, increased to the same level as that in 50% HF animals.

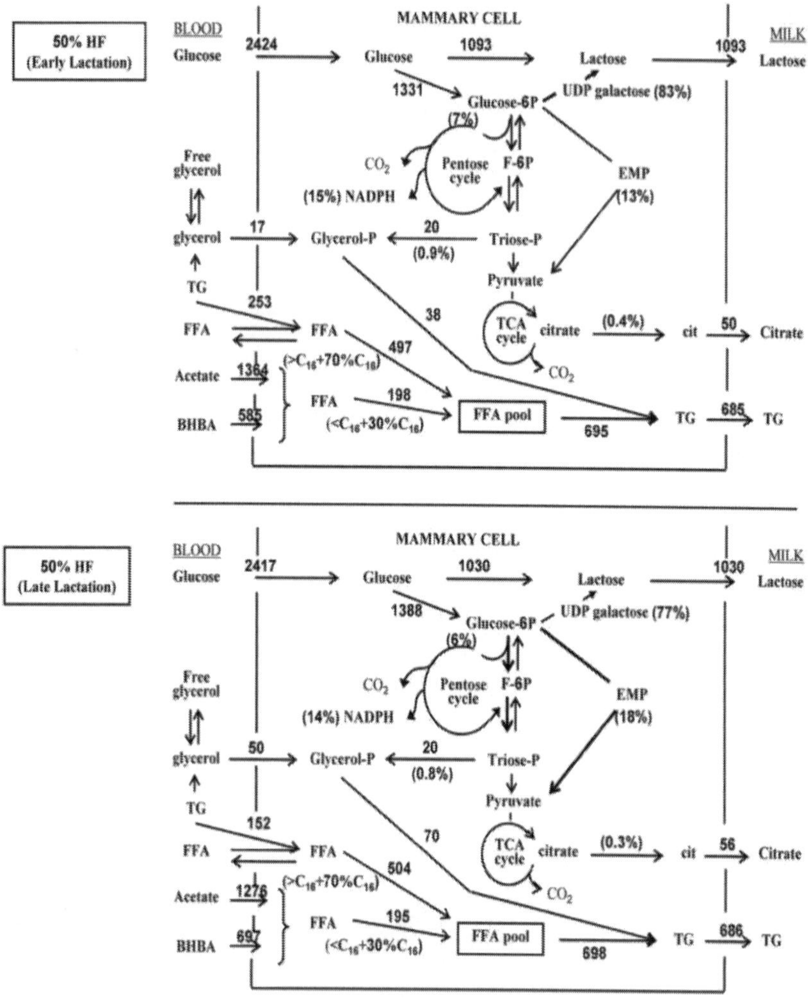

Figure 6. Scheme for calculated metabolic pathways involve intracellular partitioning of precursors of milk in mammary epithelial cells during early and late lactation of 50%HF cattle. Values are shown in micromole/min. (Calculated data from Chaiyabutr et al 2000c)

4. ROLE OF CHANGES OF HORMONE SUPPLY

It has been realized that during late pregnancy, lactogenesis occurs concurrently with mammary development and many hormones are needed for maximal stimulation of lactogenesis. There are many factors have been reported to capable of influencing lactation persistency in ruminant. The hormonal control of substrates uptake of the mammary gland and milk yield may be expected to occur in crossbred lactating cattle. However, the way in which hormones act is complex. The studies ofChaiyabutr et al (2000d) in measurements plasma hormone levels of different types of lactating crossbred cattle are shown in Table 1. There were no differences in the mean plasma thyroxine (T_4) concentrations during late pregnant period and lactating periods of experimental animals between 87.5%HF and 50%HF cattle which were given the same period and identical rations. However, plasma triiodothyronine (T_3) concentrations of both types of crossbred HF cattle were lowered in late pregnancy as compared to lactating periods. The difference of the pattern of changes between T_3 and T_4 at the onset of lactation may be suggestive of an active and rapid transformation of T_4 to T_3(Boonnamsiri et al 1979). The maintenance in high levels of plasma T_3 concentrations throughout lactation suggested that T_3 might act as an important factor in the regulation of lactation.

Changes in endocrine status during the transition period from late pregnancy to lactation will influence metabolism and the nutritional status. The pattern of differences in insulin concentrations between late pregnancy and early lactation could not be attributed to diurnal variation (Bines et al., 1983) and feeding effect (Bassett 1974). The mean plasma insulin concentration of both 87.5%HF and 50%HF cattle increased during the lactating period as compared to the late pregnant and it remained constant in a higher level throughout the lactating period. An elevation of plasma insulin levels during the onset of lactation might be a factor involved in changes in glucose turnover rate (Chaiyabutr et al., 1998). However, the plasma glucose concentrations in 87.5%HF and 50%HF cattle did not coincide with an increase in the plasma insulin concentration throughout periods of lactation. The plasma glucose concentrations of 50%HF cattle in both pregnant and lactating periods were higher than those of 87.5%HF cattle. The hyperactivity of adrenal cortex to produce the higher levels of plasma cortisol in 50%HF cattle might play a role for a rise of the plasma glucose level.

There is evidence that a lipogenic role will be expected for an elevation of plasma insulin levels during lactation by the documented fall in plasma FFA concentrations which occurred throughout lactation in crossbred HF cattle. During late pregnancy, low levels of the plasma insulin concentration coincided with higher level of plasma FFA concentrations. Several mechanisms can propose to contribute to the changes in lipogenesis, e.g. the movement of energy substrates away from the adipose tissues stores causing an elevation of plasma FFA during low levels of the plasma insulin (Yang & Baldwin, 1973) or the decreases in the sensitivity of adipose tissue to insulin at the onset of lactation (Faulkner & Pollock, 1990). During late pregnancy, mammary growth and foetus development could account for energy deficit relating to the elevation of plasma FFA concentrations. The depressed plasma insulin during late pregnancy might represent a part of the mechanism permitting mobilization of energy store. The higher level of plasma progesterone during pregnancy in both types of crossbred HF might be the other factors that contributed to an increase in plasma FFA concentration (Shevah et al., 1975). A marked elevation of plasma estradiol, primarily estrone, during late pregnancy and dramatic decreases after parturition in both 87.5%HF and 50%HF cattle, would decrease feed intake at the period of prepartum and developing high plasma FFA levels (Grummer et al., 1990).

The higher levels in plasma growth hormone (GH) during late pregnancy in 50%HF cattle could account for an increase in plasma FFA concentrations in comparison to those of 87.5%HF cattle. During lactation advance to mid and late lactation in 87.5%HF cattle, the decrease in the level of GH coincided with the decreases in the plasma FFA concentration and milk yield. The higher level of GH in early lactation of both 87.5%HF animals might influence on mammary blood flow in this period. An increase in mammary blood flow will relate to increase milk yield by contributing to a partitioning of nutrients to the mammary gland (Davis & Collier 1985; Peel et al., 1981). The control mechanism for the mammary function during transition period from pregnancy to lactation probably differed between 50%HF and 87.5%HF cattle. The higher level of GH during late pregnancy comparing to early lactation of 50%HF appeared to have no effect on mammary blood flow. The triggering of mammary blood flow and lactogenesis will involve a complex interaction of hormonal events. In late pregnancy the onset of copious lactation is overcome by the inhibition action of progesterone. Falling concentrations of progesterone after parturition would release the mammary gland from this inhibition and the rate of milk synthesis and mammary blood flow rise to a value that becomes limited by new factors perhaps the actions of GH. Plasma prolactin concentrations vary within narrow limits and did not differ between 50%HF and 87.5%HF cattle either in late pregnancy or during lactation. It has been shown that circulating prolactin in cattle can be raised or lowered by day length (Bourne &Tucker 1975) or ambient temperature (Wetteman and Tucker, 1974).

Table 1. Concentrations of triiodothyronine, thyroxine, prolactin, growth hormone, Cortisol, insulin, glucagon, progesterone and estradiol in plasma of crossbred HF animals during late pregnancy and different stages of lactation. (From Chaiyabutr et al., 2000d)

Hormones	Cattle	Late pregnant	Early lactation	Mid lactation	Late lactation
Triiodothyronine	87.5%HF	87.8	100.5	114.8	114.8
(ng/100 ml)	50%HF	87.5	132.2	131.5	126.4
Thyroxine (T4)	87.5%HF	3.66	3.67	3.94	3.69
(μg/100 ml)	50%HF	3.78	3.99	3.69	3.71
Prolactin (ng/ml)	87.5%HF	4.38	5.31	6.10	8.51
	50%HF	6.72	6.69	6.74	7.79
Growth hormone	87.5%HF	8.45	12.05	9.10	7.91
(ng/ml)	50%HF	15.16	9.03	8.76	9.76
Cortisol (ng/ml)	87.5%HF	8.9	14.2	5.7	6.2
	50%HF	6.8	29.2	25.3	35.9
Insulin	87.5%HF	12.35	18.65	23.71	21.06
(μ U/ml)	50%HF	16.55	23.03	22.91	22.92
Glucagon	87.5%HF	41.8	54.7	82.7	65.5
(pg/ml)	50%HF	57.0	67.2	118.6	112.1
Progesterone	87.5%HF	3.85	0.38	0.72	2.27
(ng/ml)	50%HF	4.45	0.12	1.42	2.16
Estradiol	87.5%HF	122.4	14.0	11.9	14.2
(pg/ml)	50%HF	175.6	18.3	22.1	25.9

5. OTHER PHYSIOLOGICAL RESPONSES IN DIFFERENT STAGES OF LACTATION IN CROSSBRED HOLSTEIN CATTLE

The mechanisms that limit the rate of milk yield and shorter lactation persistency as lactation advances in crossbred dairy cattle in tropics are not only animal genetics that have to be considered but other factors, for example, high environmental temperatures and hormonal factors can influence milk production of cows (Collier et al., 1982). High environmental temperatures are known to affect milk secretion at various levels of mechanisms in dairy cattle both directly and indirectly. Thermal effect with heat stress will affect bodily functions of dairy cattle. Many technologies are required to improve milk production of dairy cattle in the tropics. Many studies have been done in attempting to improve dairy productivity by management strategies in high environmental temperatures. Environmental modification is the most common to reduce the impacts of high temperature for increase milk production, for example water spray with fans (Fike et al., 2002), or evaporative cooling system (Chan et al., 1997; Chaiyabutr et al., 2008). Bovine growth hormone or somatotropin (bST) is a homeorrhetic hormone connected with growth and lactation in ruminant (Bauman, 1992). The concentration of plasma bovine somatotropin (bST) in 87.5% HF cattle decreased rapidly as lactation progressed to mid and late lactation(Chaiyabutr et al., 2000d). This decrease would accompany with a reduction in both mammary blood flow and milk yield. Many studies have demonstrated the efficacy of bST for improvement in milk yield (Breier et al., 1991 ; Burton et al., 1994). Long term exogenous recombinant bovine somatotropin (rbST) in 87.5% crossbred Holstein cattle increased in milk yield which accompanied with an increase in the rate of mammary blood flow, but the stimulant effect for milk yield was less in late lactation despite a high level of mammary blood flow (Chaiyabutr et al. 2007). It is not known which factors are the cause and which factors are the effects for such reduction.

The additive effects of cooling and supplemental recombinant bovine somatotropin (rbST) in responsible for the short persistency of milk yield in crossbred Holstein cattle has been carried out in 87.5%HF cattle. Long term exogenous recombinant bovine somatotropin (rbST) in 87.5% HF cattle could increase the rate of mammary blood flow, total body water (TBW) and extracellular fluid (ECF) in association with an increase in plasma IGF-I (Chaiyabutr et al. 2005). An increase in mammary blood flow would contribute nutrients partitioning to the mammary gland for milk synthesis. Greater water retention during rbST administration would not only provide a greater reservoir of soluble metabolites for biosynthesis of milk, but it may be useful in slowing down the elevation in body temperature during heat exposure.

The milk yield of 87.5% HF cattle either under misty fan cooling or non-cooling increase after peak during supplemental rbST in early stage of lactation; thereafter milk yields continued to decline as lactation advanced, while mammary blood flow and body fluids volume increased during rbST supplementation in each stage of lactation in both cooled cow under misty fan and non-cooled cow (Sitprija et al.,2010). An increase in ECF by the effects of rbST in both cooled and non-cooled cows lead to an increase in MBF as secondary responses in facilitating increased milk production. However, changes in other bodily functions of crossbred cattle were also involved by effects of mist-fan cooling and supplemental rbST. In crossbred lactating cows, the response to rbST supplementation for milk production could be enhanced under mist-fan cooling. An increase milk production during rbST supplementation was mediated via increase in efficiency of feed utilization without changes in diets digestibility (Chanchai et al., 2010). Alterations of plasma hormones concentration especially IGF-I would increase during rbST supplementation in each stage of lactation in both cooled and non-cooled cows. The effect of exogenous rbST for increase in milk productions required IGF-I as a mediator to increase in mammary blood flow for increasing the availability of substrates to the mammary gland for milk synthesis.

6. CONCLUSION

The aim of this review is to highlight the regulatory mechanisms underlying marked low milk yields, by providing an updated summary of the results obtained concerning extra-mammary factors and intra-mammary factors of crossbred dairy cows in the tropic. In conclusion, the 87.5%HF animal had the genetic potential for a high milk yield and homeorhetic adaptation for mammary function differed from 50%HF animals during periods of lactation. Altering lactation persistency in 87.5%HF was regulated mainly by chronically acting of growth hormones through the period of lactation. The utilization of glucose in the mammary gland was determined by measuring rates of glucose uptake and the incorporation of glucose into milk components in both groups of 50% HF and 87.5% HF cattle. In early lactation, there were no significant differences of the total glucose entry rate and glucose carbon recycling between crossbred cattle. The percentages and values of non-mammary glucose utilization increased during lactation advance in both 50% HF and 87.5% HF cattle. The percentage of glucose uptake for utilization in the synthesis of milk lactose by the mammary gland of 87.5% HF cattle was higher than 50% HF cattle. Intracellular glucose6-phosphate metabolized via the pentose phosphate pathway accounted for the NADPH (reducing equivalent) of fatty acid synthesis in the mammary gland being higher in 87.5% HF animals during mid- lactation. A large proportion of metabolism of glucose via the Emden-Meyerhof pathway in the mammary gland was more apparent in 50% HF cattle than 87.5% HF cattle during early and mid- lactation, while it markedly increased for 87.5% HF cattle during late lactation. The glucose utilization for biosynthetic pathways in the mammary gland of 50% HF animals was maintained in a similar pattern throughout periods of lactation. A poorer lactation persistency of 87.5% HF animals occurred during lactation advance, which was related to a decrease in the lactose biosynthetic pathway. As advanced lactation, local changes for biosynthetic capacity within the mammary gland would be a factor in identification of the utilization of substrates in the rate of decline in milk yield. The proportion of glucose would be metabolized less for lactose synthesis, but metabolized more via the Embden-Meyerhof pathway and the tricarboxylic acid cycle as lactation advances to late lactation.

REFERENCES

1. E. F. Annison, J. L. Linzell, C. E. West, 1968Mammary and whole animal metabolism of glucose and fatty acids in fasting lactating goats. Journal of Physiology, 197445459

2. Bassett J.M.1974Diurnal patterns of plasma insulin, GH, corticosteriod and metabolite concentrations in fed and fasted sheep.Australian Journal of Biological Sciences, 27167168

3. Bauman D.E. & Currie W.B.1980Partitioning of nutrients during pregnancy and lactation: a review of mechanisms involving homeostasis and homeorhesis. Journal of Dairy Science, 6315141529

4. Bauman D.E.1992Bovine somatotropin: review of an emerging animal technology. Journal of Dairy Science, 75334323451

5. R. Bickerstaffe, E. F. Annison, J. L. Linzell, 1974The metabolism of glucose, acetate, lipids and amino acids in lactating dairy cows. Journal of Agricultural Science, 827185

6. J. A. Bines, I. C. Hart, S. V. Morant, 1983Endocrine control of energy metabolism in the cow: diurnal variations in the concentrations of hormones and metabolites in the blood plasma of beef and dairy cows. Hormone and metabolic Research, 15330334

7. V. Boonnamsiri, J. C. Kermode, B. D. Thompson, 1979Prolonged intravenous infusion of labelled iodocompounds in the rat: 125I thyroxine and (125I) triiodothyronine metabolism and extrathyroidol conversion of thyroxine to triiodothyronine. Journal of Endocrinology, 82235243

8. R. A. Bourne, H. A. Tucker, 1975Serum prolactin and LH responses to photoperiod in bull calves. Endocrinology, 97473475

9. B. H. Breier, P. D. Gluckman, S. N. Mc Cutcheon, S. R. Davis, 1991Physiological responses to somatotropin in the ruminant. Journal of Dairy Science, 74suppl.2), 2034

10. A. F. Burnol, A. Leturque, M. Loizeaau, C. Postic, J. Girard, 1990Glucose transporter expression in rat mammary gland. Biochemical Journal, 270277279

11. J. L. Burton, B. W. Mc Bride, E. Block, D. R. Glimm, J. J. Kennelly, (1994, 1994A review of bovine growth hormone. Canadian Journal of Animal Science, 74167201

12. N. Chaiyabutr, A. Faulkner, M. Peaker, 1980The utilization of glucose for the synthesis of milk components in the fed and starved lactating goat in vivo. Biochemical Journal, 186301308

13. N. Chaiyabutr, S. Komolvanich, S. Sawangkoon, S. Preuksagoon, S. Chanpongsang, 1997The regulation of body fluids and mammary circulation during late pregnancy and early lactation of crossbred Holstein cattle feeding on different types of roughage. Journal of Animal Physiology and Animal Nutrition, 77167179

14. N. Chaiyabutr, S. Komolvanich, S. Sawangkoon, S. Preuksagorn, S. Chanpongsang, 1998Glucose metabolism in vivo in crossbred Holstein cattle feeding on different types of roughage during late pregnancy and early lactation. Comparative Biochemistry and Physiology. Part A, 119905913

15. N. Chaiyabutr, S. Preuksagorn, S. Komolvanich, S. Chanpongsang, 2000aComparative study on the regulation of body fluids and mammary circulation at different states of lactation in crossbred Holstein cattle feeding on different types of roughage. Journal of Animal Physiology and Animal Nutrition, 837484

16. N. Chaiyabutr, S. Preuksagorn, S. Komolvanich, S. Chanpongsang, 2000bGlucose metabolism in crossbred Holstein cattle feeding on two types of roughage at different stages of lactation. Comparative Biochemistry and Physiology. Part A, 125121130

17. N. Chaiyabutr, S. Komolvanich, S. Preuksagorn, S. Chanpongsang, 2000cComparative studies on the utilization of glucose in the mammary gland of crossbred Holstein cattle feeding on different types of roughage during different stages of lactation. Asian-Australasian Journal of Animal Science, 13334347

18. N. Chaiyabutr, S. Komolvanich, S. Preuksagorn, S. Chanpongsang, 2000dPlasma levels of hormones and metabolites as affected by the forages type in two different types of crossbred Holstein cattle. Asian-Australasian Journal of Animal Science, 1313591366

19. N. Chaiyabutr, S. Thammacharoen, S. Komolvanich, S. Chanpongsang, 2002Studies on the mode of uptake of plasma glucose, acetate, β-hydroxybutyrate triglyceride fatty acids and glycerol by the mammary gland of crossbred Holstein cattle feeding on different types of roughage. Asian-Australasian Journal of Animal Science, 1514451452

20. N. Chaiyabutr, S. Thammacharoen, S. Komolvanich, S. Chanpongsang, 2005Effects of long-term administration of recombinant bovine somatotropin on milk production and insulin like growth factor-I and

insulin in crossbred Holstein cows. Journal of Agricultural Science (Cambridge), 143311318

21. N.. Chaiyabutr, S. Thammacharoen, S. Komolvanich, S. Shanpongsang, 2007Effects of long-term exogenous bovine somatotropin on water metabolism and milk yield in crossbred Holstein cattle. Journal of Agricultural Science (Cambridge), 145173184
22. N. Chaiyabutr, S. Chanpongsang, S. Suadsong, 2008Effects of evaporative cooling on the regulation of body water and milk production in crossbred Holstein cattle in a tropical environment. International Journal of Biometeorology, 52575585
23. S. C. Chan, J. T. Hubber, K. H. Chen, J. M. Simas, Z. Wu, 1997Effects of ruminally inert fat and evaporative cooling on dairy cows in hot environmental temperatures. Journal of Dairy Science, 8011721178
24. W. Chanchai, S. Chanpongsang, N. Chaiyabutr, 2010Effects of cooling and supplemental recombinant bovine somatotropin on diet digestibility, digestion kinetics and milk production of cross-bred Holstein cattle in the tropics. Journal of Agricultural Science (Cambridge), 148233242
25. L. E. Chase, 1993Developing nutrition programs for high producing dairy herds. Journal of Dairy Science, 7632873293
26. R. J. Collier, D. K. Beede, W. W. Thatcher, L. A. Israel, C. J. Wilcox, 1982Influences of environment and its modification on dairy animal health and production. Journal of Dairy Science, 6522132227
27. J. B. Coulon, L. Perochon, F. Lescourret, 1995Modelling the effect of the stage of pregnancy on dairy cow's milk yield. Journal of Animal Science, 60401408
28. C. L. Davis, D. E. Bauman, 1974General metabolism associated with the synthesis of milk. In: Lactation Vol. II, B.L. Larson,& V.R Smith, (Eds.), 330Academic Press, New York and London.
29. Davis S.R, & Collier R.J.1985Mammary blood flow and regulation of substrate supply for milk synthesis. Journal of Dairy Science, 6810411058
30. K. E. Ebner, F. L. Schanbacher, 1974Biochemistry of lactose and related carbohydrates. In: Lactation Vol. II,. B.L. Larson,& V.R Smith, (Eds.), 77113Academic Press, New York and London.
31. A. Faulkner, M. Peaker, 1987Regulation of mammary glucose metabolism in lactation. In: The Mammary gland: development, regulation and function, M.C. Neville & Daniel CW (Eds.), 535562Plenum Press, New York.

32. A. Faulkner, H. T. Pollock, 1990Metabolic responses to euglycaemic hyperinsulinaemia in lactating and non-lactating sheep in vivo. Journal of Endocrinology, 1245966

33. J. H. Fike, C. R. Staples, L. E. Sollenberger, J. E. Moore, H. H. Head, 2002Southeastern pasture-based dairy systems: housing, posilac, and supplemental silage effects on cow performance. Journal of Dairy Science, 85866878

34. R. R. Grummer, S. J. Bertics, D. W. La Count, J. A. Snow, M. R. Dentine, R. H. Staauffacher, 1990Estrogen induction of fatty liver in dairy cattle. Journal of Dairy Science, 731537

35. G. L. Hahn, T. L. Mader, J. B. Gaughan, Q. Hu, J. A. Nienaber, 1999Heat waves and their impacts on feedlot cattle, Proceedings of 15th International Society Biometerology congress, Sydney, Australia, September, 353357

36. A. Hanwell, M. Peaker, 1977Physiological effects of lactation on the mother. In: Comparative Aspects of Lactation, M. Peaker, (Ed.), 297312The Zoological Society of London, Academic Press.

37. D. C. Hardwick, J. L. Linzell, S. M. Price, 1961The effect of glucose and acetate on milk secretion by the perfused goat udder. Biochemical Journal, 803745

38. D. C. Hardwick, J. L. Linzell, T. M. Mepham, 1963The metabolism of acetate and glucose by the isolated perfused udder 2. The contribution of acetate andglucose to carbondioxide and milk constituents. Biochemical Journal, 88213220

39. I. C. Hart, 1973Effect of 2-Bromo-2-ergocryptine on milk yield and the level of prolactin and growth hormone in the blood of the goat at milking. Journal of Endocrinology. 57179180

40. I. C. Hart, S. E. Lawrence, T. B. Mepham, 1980Effect of exogenous growth hormone on mammary blood flow and milk yield in lactating goats. Journal of Physiology,46308

41. S. Horsfield, J. M. Infield, E. F. Annison, 1974Compartmental analysis and model building in the study of glucose kinetics in the lactation cow. Proceedings of the Nutrition Society, 331915

42. M. C. N. Jayasuriya, H. G. D. Perera, 1982Urea-ammonia treatment of rice straw to improve its nutritive value for ruminants. Agricultural Wastes, 4143150

43. E. A. Jones, 1978Lactose Biosynthesis. In: Lactation Vol. II, B.L. Larson,& V.R Smith, (Eds.), 371385Academic Press, New York and London.

44. T. Komatsu, F. Itoh, S. Kushibiki, K. Hodate, 2005Changes in gene expression of glucose transporters in lactating and nonlactating cows. Journal of Animal Science, 83557564

45. W. Kaufmann, H. Hagemeister, 1987Composition of milk. In: Dairy-cattle production, H.O. Gravert (Ed), 107171Elsevier Science Publishers BV, Amsterdam.

46. N. J. Kuhn, D. T. Carrick, C. J. Wilde, 1980Milk synthesis. Journal of Dairy Science, 63328336

47. P. Lacasse, V. C. . Farr, S. R. Davis, C. G. Prossser, 1996Local secretion of nitric oxide and the control of mammary blood flow. Journal of Dairy Science, 7913691374

48. R. A. Leng, C. E. West, 1969Contribution of acetate, butyrate, palmitate, stearate and oleate to ketone body synthesis in sheep. Research in Veterinary Science, 105763

49. D. B. Lindsay, 1973Metabolic changes induced by pregnancy in the ewe. In: Production disease in farm animals, E.J.M. Payne; K.G. Hibitt & B.F. Sansom, (Eds). 107114Bailliere, Tindal, London.

50. J. L. Linzell, T. B. Mepham, 1968Mammary synthesis of amino acids in the lactating goat. Biochemical Journal, 1071819

51. J. L. Linzell, M. Peaker, 1971Mechanisms of milk secretion. Physiological Review, 51564597

52. J. L. Linzell, 1968The magnitude and mechanisms of the uptake of milk precursors by the mammary gland. Proceedings of the Nutrition Society,274452

53. Linzell J.L.1973The demands of the udder and adaptation to lactation. In: Production disease in farm animals,J.M. Payne; K.G. Hibbitt & B.F. Sansom (Eds.),89106Bailliere, Tidal, London.

54. J. L. Linzell, T. B. Mepham, 1974Effect of intramammary arterial infusion of essential amino acids in the lactating goat. Journal of Dairy Science, 41101109

55. J. L. Linzell, 1974Mammary blood flow and methods of identifying and measuring precursors of milk. In: Lactation Vol. II, B.L. Larson,& V.R Smith, (Eds.), 143225Academic Press, New York and London.

56. Farlane. M. V. a. Mac, B. Howard, 1970Water in the physiological ecology of ruminants. In: Physiology of digestion and metabolism in Ruminants. Phillipson, A.T. (Ed.), 362374Oriel Press, Newcastle upon Tyne.

57. T. B. Mepham, 1993The development of ideas on the role of glucose in regulating milk secretion. Australian Journal of Agricultural Research, 44508522

58. M. R. Murphy, 1992Symposium: Nutritional factors affecting animal water and waste quality. Journal of Dairy Science, 75326333

59. R. M. Nakamura, C. T. Araki, N. Chaiyabutr, 1993Temperate dairy cattle for hot climates: Telemetry studies and strategy. In: Livestock Environment IV, Fourth International Symposium, 1622University of Warwick, England.

60. D. L. Palmquist, C. L. Davis, R. E. Brown, D. S. Sachan, 1969Availability and metabolism of various substrates in ruminants. V. Entry rate into the body and incorporation into milk fat of D(-)-β-hydroxybutyrate. Journal of Dairy Science, 52633638

61. C. J. D. Peel, D. E. Bauman, R. C. Gorwit, C. J. Sniffen, 1981Effect of exogenous growth hormone on lactational performance in high yielding dairy cows. Journal of Nutrition, 11116621671

62. S. Promma, I. Tasaki, B. Cheva-Isarakul, T. Indratula, 1994Digestibility of Neutralized urea-treated rice straw and nitrogen retained in crossbred Holstein streers. Asian-Australasian Journal of Animal Science, 7487491

63. S. K. Ranjhan, I. A. P. Kalanidh, T. K. Gosh, U. B. Singh, K. K. Saxena, 1982Body composition and water metabolism in tropical ruminants using tritiated water. In: Use of Tritiated Water in Studies of Production and Adaptation in Ruminants. 117132International Atomic Energy Agency, Vienna.

64. R. L. Reid, N. T. Hinks, 1962Studies on the carbohydrate metabolism of sheep, XVIII. The metabolism of glucose, free fatty acid, ketones and amino acids in late pregnancy and lactation. Australian Journal of Agricultural Research, 1311121123

65. J. A. F. Rook, J. V. Wheelock, 1967The secretion of water and water soluble constituents in milk. Journal of Dairy Research. 34273287

66. Y. Shevah, W. J. M. Black, W. R. Carr, R. B. Land, 1975The effects of nutrition on the reproductive performance of Finnx Dorset ewes. 1. Plasma progesterone and LH concentrations during late pregnancy. Journal of Reproductive Fertility, 45283288

67. J. E. Shirley, R. S. Emerry, E. M. Convey, W. D. Oxender, 1973Enzymic changes in bovine adipose and mammary tissue, serum and mammary tissue hormonal changes with initiation of lactation. Journal of Dairy Research, 56569574

68. S. Sitprija, S. Chanpongsang, N. Chaiyabutr, 2010Effects of Cooling and Recombinant Bovine Somatotropin Supplementation on Body Fluids, Mammary Blood Flow, and Nutrients Uptake by the Mammary Gland in Different Stages of Lactation of Crossbred Holstein Cattle. Thai Journal of Veterinary Medicine, 40195206

69. J. Sölkner, W. Fuchs, 1987A comparison of different measures of persistency with special respect to variation of test-day milk yields. Livestock Production Science 16305319

70. G. E. Thompson, E. M. Thomson, 1977Effect of cold exposure on mammary circulation oxygen consumption and milk secretion in the goat. Journal of Physiology 272187196

71. H. A. Tucker, 1981Physiological control of mammary growth, lactogenesis and lactation. Journal of Dairy Science, 6414031421

72. H. A. Tucker, (1987). Quantitative estimates of mammary growth during various physiological states: a review. Journal of Dairy Science, Vol. 70, pp.1958-

73. C. E. West, E. F. Annison, J. L. Linzell, 1967Plasma free fatty acid uptake and release by the goat mammary gland. Biochemical Journal, 102P.

74. C. E. West, R. Bickerstaffe, E. F. Annison, J. L. Linzell, 1972Studies on the mode of uptake of blood triglycerides by the mammary gland of the lactating goat. The uptake and incorporation into milk fat and mammary lymph on labelled glycerol, fatty acids and triglycerides. Biochemical Journal, 126477490

75. R. P. Wetteman, H. A. Tucker, 1974Relationship of ambient temperature to serum prolactin in heifers. Proceedings Society of Experimental Biology and Medicine, 146908911

76. H. G. Wood, G. J. Peeters, R. Verbeke, M. Lauryssens, B. Jacobson, 1965Estimation of the pentose cycle in the perfused cow's udder. Biochemical Journal, 96607615

77. C. Xiao, J. P. Cant, (2003, 2003Glucose transporter in bovine mammary epithelial cells is an asymmetric carrier that exhibits cooperativity and trans-stimulation. American Journal of Physiology, Cell Physiolology, 285C1226C1234

78. Y. T. Yang, R. L. Baldwin, 1973Lipolysis in isolated cow adipose cells. Journal of Dairy Science, 56366374

79. F. Q. Zhao, W. M. Moseley, H. A. Tucker, J. J. Kennelly, 1996Regulation of glucose transporter gene expression in mammary gland, muscle, and fat of lactating cows by administration of bovine growth hormone and bovine growth hormone-releasing factor. Journal of Animal Science, 74183189

CHAPTER 6

Occurrence of *Staphylococcus aureus* on Farms with Small Scale Production of Raw Milk Cheeses in Poland

Jolanta G. Rola *, Anna Czubkowska, Weronika Korpysa-Dzirba and Jacek Osek

Department of Hygiene of Food of Animal Origin, National Veterinary Research Institute, 24-100 Pulawy, Poland

ABSTRACT

This paper describes the results of a 3-year study on the prevalence, enterotoxinogenicity and resistance to antimicrobials of *S. aureus* isolated on dairy farms with small scale production of raw cow milk cheeses. The samples of raw milk, semi-finished products and the final products as well as swabs were collected between 2011 and 2013 from nine dairy farms in Poland. A total of 244 samples were examined, of which 122 (50.0%) were contaminated with *S. aureus* including 18 of 26 (69.2%) mature cheese samples with log10 CFU g^{-1} between <1- and 7.41. In swabs collected from the staff and production environment the highest contamination rate with coagulase positive staphylococci (CPS) was detected on hands of cheese makers (4.34 log10 CFU/swab). None of the cheese samples contaminated with CPS contained staphylococcal enterotoxins (SEs). However, 55 of 122 (45.1%) *S. aureus* isolates possessed SEs genes, mainly (26 of 55; 47.3%) a combination of the *sed*, *sej* and *ser* genes. Furthermore, the *sep* (15 of 55; 27.3%) as well as *seg* and *sei* (9 of 55; 16.4%) genes were also identified. The remaining *S. aureus* isolates possessed the *sea* gene (one isolate), the combination of *sec*, *seg* and *sei* (three isolates) as well as the *sed*, *sej*, *sep* and *ser* markers together (one CPS). Resistance to penicillin (62 of 122 isolates; 50.8%) was the most common among the tested isolates. Some CPS were also resistant to chloramphenicol (7; 5.7%) and tetracycline (5; 4.1%). The obtained results indicated that the analyzed cheeses were safe for consumers. To improve the microbiological quality of traditional cheese products more attention should be paid to animal welfare and hygiene practices during the process of cheese manufacturing in some dairy farms.

KEYWORDS

S. aureus; SE genes; antimicrobial resistance; raw milk cheese; public health

1. INTRODUCTION

Staphylococcal food poisoning (SFP) is one of the most common foodborne diseases worldwide resulting from the consumption of foods containing staphylococcal enterotoxins (SEs) produced mainly by *Staphylococcus aureus* [1,2]. Symptoms of SFP have a rapid onset (2–8 h) and include hypersalivation, nausea, vomiting, abdominal cramping and diarrhea [3]. In the majority of cases recovery occurs within 24–48 h without specific treatment. Occasionally, the disease can be more severe or even fatal, especially in infants, elderly or immunecompromised patients. *S. aureus* is ubiquitous in the environment and can be found in the air, water, humans and animals. It is also one of the major causes of bovine mastitis and therefore, raw milk and subsequently raw milk products may be contaminated with *S. aureus* [4]. About 10% of cheeses in Europe are made from raw milk, which presents a considerable potential threat to human health [5]. In Scotland, *S. aureus*was found to be the most frequent pathogen of raw milk cheeses [6]. In France, a study of foodborne disease outbreaks showed that *S. aureus* was one of the most common causative pathogens associated with milk-related outbreaks [7]. The lack of proper hygienic measures during food processing may also increase the probability of contamination with *S. aureus*, especially in manually prepared foods. Cheese makers carrying enterotoxin-producing *S. aureus* in their noses or on their hands are thought to be the main source of food contamination caused byphysicial contact or through respiratory secretions. André *et al.* isolated *S. aureus* from hand and nose samples of approximately 75% of cheese makers in a dairy processing plant [8].

Some *S. aureus* produce toxins which are potent emetic agents causing SFP. In Italy, 55% of food isolates (milk, dairy products, meat, meat products) were positive for classic SEs (SEA-SEE) [9], while in Norway 48% of isolates from bovine raw milk and raw milk products were found to be SE producers [10]. Many *S. aureus* strains can synthesize more than one type of toxins [1,2] and to date 22 different SEs have been described. Apart from the classic SE types several new variants of SEs or staphylococcal-like toxins have been identified [11]. All of these toxins are heat stable and therefore, they may still be present in food while *S. aureus* is absent [3]. Enterotoxinogenic staphylococci need to reach levels of at least 5-6 log10 CFU g^{-1} to produce detectable amounts of SE [12]. More than 95% of SFP outbreaks worldwide are caused by classic enterotoxins, mainly SEA. On the other hand, strains of *S. aureus* isolated from cow milk are mostly positive for SEC and SED [12].

According to the recent European Food Safety Authority report, a total of 386 food-borne outbreaks caused by staphylococcal toxins were identified in the European Union in 2013 [13]. Most of them (336 outbreaks) were recorded in France where consumption of unpasteurized milk cheeses is common and milk-based products are more frequently involved in food poisoning than in other countries. In Poland, in the same year, only five confirmed food-borne intoxications were notified [13]. In recent years, traditionally made food products such as cheese manufactured at farm dairies are becoming more popular in our country [14]. These products should be of high quality and safe for consumers.

The aim of the study was to determine the occurrence of *S. aureus* in dairy farms with small scale production of raw cow milk cheese. Moreover, enterotoxinogenicity and resistance to antimicrobials of the isolates were investigated.

2. RESULTS

CPS were isolated from 122 (50.0%) out of 244 tested samples (Table 1). The bacteria were detected both in raw milk (12; 46.2%) and in the final cheese products (18; 69.2%). The highest percentage of CPS positive samples was found in the formed cheese (21 of 26; 80.8%), in grains after rinsing (19 of 25; 76.0%), milk curd (19 of 27; 70.4%) and in milk after heating (8 of 12; 66.7%). Of the 25 positive swabs CPS were most frequently isolated from the hands of cheese makers (11 of 26; 42.3%) and milk tanks (7 of 26; 26.9%) (Table 1).

Table 1. Contamination with coagulase positive staphylococci (CPS) at the different stages of cheese production.

Collected Material	Source	No. of Samples	No. (%) of CPS Positive Samples	Ranges of CPS at Different Production Stages								
				Farm No.								
				1	2	3	4	5	6	7	8	9
Swab				Log10 CFU/swab or CFU/cm^2								
	Hands of cheese maker	26	11 (42.3)	0	0	0–4.34	0	0–3.23	0–2.72	0–2.70	0–3.81	0–3.12
	Milk tank	26	7 (26.9)	0	0	0.24–1.70	0	0–0.18	0	0–0.30	0–0.81	0–0.30
	Strainer/sacks	26	3 (11.5)	0	0	0–2.83	0	0	0–2.00	0	0	0
	Cheese mould	24	4 (16.7)	0–1.40	0	0–2.54	0	0–2.10	0	0	0	0–2.00
Sample				Log10 CFU mL^{-1} or CFU g^{-1}								
	Raw milk	26	12 (46.2)	0	0	0–3.63	0–1.18	0–4.74	0–2.63	0	0–5.00	1.08–3.04
	Milk after heat treatment	12	8 (66.7)	0–1.70	0–0.18	0.70–1.48	1.18	2.36	1.73	0	0	3.90
	Curd	27	19 (70.4)	<1–2.64	2.18–2.53	<1–4.77	<1–1.60	<1–4.65	2.89–4.89	<1–3.83	<1–5.83	1–4.26
	Grains after rinsing	25	19 (76.0)	<1–3.36	<1–2.57	2.61–5.92	1.30–2.04	1.74–5.15	4.08–5.32	<1–4.74	<1	3.28–5.00
	Formed cheese	26	21 (80.8)	<1–3.95	<1–2.40	3.23–5.08	1–2.46	1.30–6.04	4.53–5.63	<1–4.74	<1–5.60	3.62–5.59
	Mature cheese	26	18 (69.2)	2.60–3.82	<1–5.53	1–3.08	<1	<1–4.91	<1–5.74	2.83–6.58	<1–7.41	4.81–7.11
Total		244	122 (50.0)									

Contamination with CPS depended on the type of material (Table 1). Only on two dairy farms the number of CPS in raw milk was 0 log10 CFU mL^{-1}, whereas on other farms contamination was ranging from 0–1.0 to 0–5.0 log10 CFU mL^{-1}. All semi-finished products obtained after addition of rennet were positive for CPS at the maximum levels of <1–5.83, 2.61–5.92 and 1.30–6.04 log10 CFU g^{-1} in milk curd, grains after rinsing and formed cheese, respectively. Furthermore, the final products in all dairy farms were contaminated with staphylococci and the levels ranged from 0 to <1–7.41 log10 CFU g^{-1}. In swabs collected from production environment the highest contamination rate with CPS was detected on hands of cheese makers (4.34 log10 CFU/swab).

All CPS were coagulase- and catalase-positive. β-haemolysis was observed in 108 of 122 (88.5%) isolates. Based on the presence of the *nuc* and *16S rRNA* genes all isolates were identified as *S. aureus*. Each sample contaminated with CPS, as well as all final products (total 105 samples) were tested for the presence of staphylococcal enterotoxins (SEs) and all of them were negative. However, SE genes were found in 55 (45.1%) isolates. Most of them were positive for three enterotoxin markers, *i.e.*, *sed* + *sej* + *ser* and *sec* + *seg* + *sei* (26 and three isolates, respectively). Some CPS had only one gene—*sep* (15; 12.0%) or *sea* (1; 1.8%) or four *sed* + *sej* + *sep* + *ser* toxin genes (1; 1.8%) (Figure 1).

Detection of SEA-SEE in culture supernatants of the isolates possessing the *sea—see* genes resulted in the presence of enterotoxins in supernatants of 31 of 55 (56.4%) such isolates.

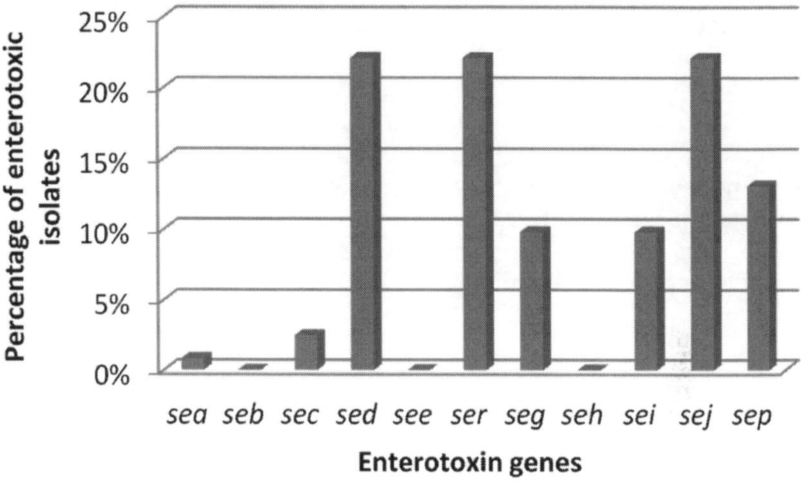

Figure 1. Enterotoxigenic *S. aureus* isolates.

Resistance to penicillin (62 of 122 isolates; 50.8%) was the most common among the tested *S. aureus*, followed by chloramphenicol (7; 5.7%), tetracycline (5; 4.1%), sulphamethoxazole (1; 0.8%) erythromycin (1; 0.8%) and streptomycin (1; 0.8%) (Figure 2). Of the 62 penicillin resistant *S. aureus* isolates, nine were also resistant to other antimicrobials: four to chloramphenicol, three to tetracycline, one to sulphamethoxazole and one to chloramphenicol together with streptomycin.

3. DISCUSSION

In this study, the prevalence and enterotoxigenicity of *S. aureus* isolated on dairy farms with small scale production of raw milk cheeses were tested. The results indicated that 46.2% of the raw milk samples were contaminated with CPS at the level up to 5 log10 CFU mL^{-1}. It is interesting that in milk from two farms, CPS were not detected. A similar level of contamination of raw milk with CPS has been found by other authors [15,16,17,18,19]. However, there are also publications in which the rate of contamination was higher than in the present study. Peles *et al.* showed that bulk tank milk from 14 out of 20 farms was contaminated with *S. aureus* at levels up to 3.78 log10 CFU mL^{-1} [20]. Tondo *et al.* reported that *S. aureus*was present in 90.4% of raw milk samples with a mean number of 3.54 log10 CFU mL^{-1} [21]. These bacteria may be introduced to bulk milk either by direct excretion from the udder of a cow with clinical or subclinical staphylococcal mastitis or by fecal contamination [22]. In well drawn milk contamination with *S. aureus* ranged from 2.0 to 2.30 log10 CFU mL^{-1} but in the case of the presence of bacteria in the udder the number of these microorganisms may increase up to 4 log10 CFU mL^{-1} [23]. In the present study neither clinical nor subclinical mastitis were indicated in the cows from the dairy farms examined. Therefore, one of the possible explanations of the higher *S. aureus* level in some raw milk samples may be their contamination from the milking equipment or personnel involved in production.

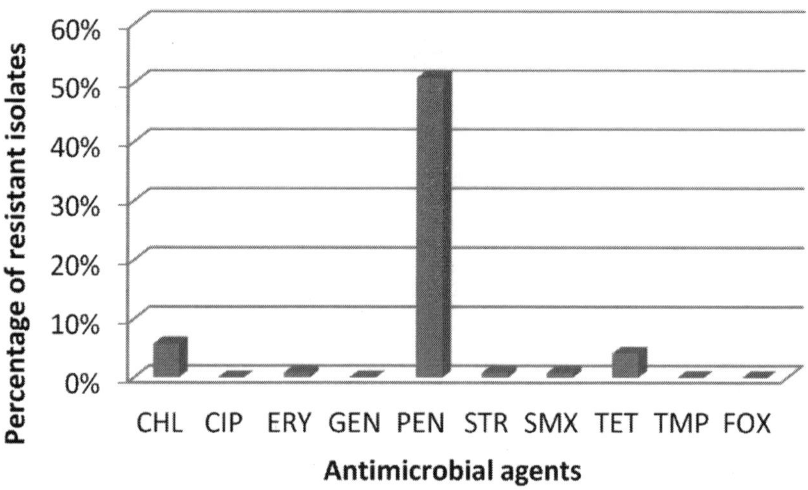

Figure 2. Antimicrobial resistance of *S. aureus* isolates.

The results of the current investigation showed that the contamination levels of the cheese gradually increased during its manufacture. It was observed that in farms in which both the quality of milk was high and the hands of cheese makers were negative for CPS, the number of staphylococci in mature cheese was lower or equal to 5 log10 CFU g^{-1}. On the other hand, in farms with a high number of CPS both in milk and in hand's swabs, the level of contamination increased along the production chain and peaked in the final products at as high as 6–7 log10

CFU g^{-1}. An increase in contamination with *S. aureus*, of traditionally made cow milk cheese has been reported by others authors, especially at the stage from milk to curd [18]. These increasing levels may be explained by physical entrapment of *S. aureus* in the curd and the ability of these bacteria to grow rapidly in milk, as demonstrated by their generation time of 0.8 h at 25 °C [24,25]. Another possible source of contamination are people involved in cheese manufacturing, since *S. aureus* is frequently found on the skin of cheese makers [8,26]. In the present study, CPS were quite often detected on the hands of cheese makers in six out of nine tested farms and this could be the main source of contamination in the later stages of cheese manufacture.

None of the cheese contaminated with CPS contained staphylococcal enterotoxins but 45.1% isolated *S. aureus* harbored the SE genes. However, enterotoxins' production by such strains is possible under appropriate environmental conditions, especially considering the temperature. The capability of the isolates for the production of enterotoxins A – E was confirmed in all *sea – see*-positive (31 of 122; 25.4%) isolates. Eight types of the SE genes (*sea*, *sec*, *sed*, *seg*, *sei*, *sej*, *sep*, *ser*) were detected in the isolated CPS and most of them (26 of 55; 47.3%) harbored the combination of the *sed*, *sej* and *ser* genes. The coexistence of the *sed* and *sej* markers has previously been reported by other authors and is due to the common location of the genes on the same plasmid [27]. The second largest subgroup of enterotoxigenic CPS detected in the present study was that with the*sep* gene (27.3%). Only one *S. aureus* strain had the *sea* gene. These results are similar to the study of Hummerjohann *et al.*who showed that *S. aureus* with the SE gene pattern *sed*, *sej* and *ser* dominated among the isolates recovered from Swiss raw milk cheeses [28]. Furthermore, several Italian studies also reported a dominance of the *sed* marker, which was often associated with the *sea* and *sej* genes [9,29,30]. In Norway, Jorgensen *et al.* indicated that 55% of CPS isolated from bovine bulk milk and 14.7% recovered from different stages of cheese production possessed SE genes but only the *seg* and *sei*markers were identified [19]. On the other hand, the present results are in contrast to the studies from Japan, Austria, France and Turkey, where the *sec* marker was commonly detected in *S. aureus* originating from milk and raw milk cheeses [10,31,32,33]. These data may possibly suggest a prevalence of certain *S. aureus* in different geographical regions.

The present study has shown that most *S. aureus* isolated from cheese were resistant to penicillin (50.8%); only a few of them showed resistance to tetracycline and sulfamethoxazole. Most of the CPS (47.5%) were resistant to one antimicrobial which is typical for strains of veterinary origin, whereas in human isolates rather multiresistant strains predominate [34]. Our results are comparable to those reported by other investigators [9,20]. The present results of antimicrobial resistance are in correlation with antibiotics that are used for treatment of bovine mastitis in Poland. Krasucka *et al.*, based on a questionnaire completed by 109 veterinarians from the whole of Poland, showed that penicillins are mostly used in treatment of cattle infections [35]. According to the report of the European Medicines Agency, in Poland in 2012 penicillins, tetracyclines and sulfonamides accounted for over 75% of the total of veterinary antimicrobial agents sold [36].

The obtained results have also indicated resistance to chloramphenicol in 5.7% of the isolates tested, although this antibiotic is banned in the European Union for animals bred for human consumption. However, this resistance may be the result of the presence of chloramphenicol naturally occurring in the environment [37].

The obtained results showed that the analyzed naturally ripened rennet cheese was safe for consumers. However, a significant proportion of samples was contaminated with *S. aureus* and therefore, more attention should be paid to animal welfare and hygiene of cheese manufacturing, especially quality of raw milk to improve the safety of these traditional products.

4. MATERIALS AND METHODS

4.1. Sample Collection

A total of 244 samples from various stages of cheese production were collected between 2011 and 2013 from nine dairy farms located in the north-eastern part of Poland. They were small dairy farms rearing 4–6 milking cows and producing naturally ripened rennet cheese from raw unpasteurized cow milk. The production of such cheeses proceeds as follows: raw milk is heated to about 32 °C, the rennin added starts the process of cloth formation. Afterwards, the milk curds are cut and the liquid whey is separated. In the next stage, cheese grains are rinsed with warm water and once again liquid whey is separated. The grains are placed in special forms where the cheese is drained off, formed and the salt is added. After this step, the cheeses are transported to the specially assigned places for ripening. The following samples were collected from cheese production stages of which the highest level of *S. aureus* was expected: 26 of raw milk, 90 of semi-finished products (heated milk, curd, grains, formed cheese), 26 of final products and 102 of swabs from the production environment. Once on each farm, 8–10 samples were taken from the same stages of the production chain. Raw milk and semi-finished products were collected on the day of cheese production, whereas the final products were taken after two weeks of cheese ripening. The swabs were obtained using sterile 50 cm^2 sampling sponges with maximum recovery diluent (TSC Ltd., Lancashire, UK). The swabs from the production environment, *i.e.*, milk tanks, were collected using 10 × 10 cm^2 sterile propylene templates (TSC).

4.2. Enumeration of CPS

Enumeration of CPS was performed on Baird-Parker agar with rabbit plasma fibrinogen (bioMerieux, Marcy-I'Etoile, France). After 48 h of incubation at 37 ± 1 °C, one typical colony was taken for further identification [38].

4.3. Identification of S. aureus

S. aureus was identified by PCR detection of the *nuc* and *16S rRNA* genes according to the protocols recommended by the European Union Reference Laboratory for Antimicrobial Resistance [39].

4.4. Detection of Staphylococcal Enterotoxins Genes

The presence of SE genes was determined using two multiplex PCR assays. The first one was performed with six pairs of primers allowing detection of the genes: *sea*, *seb*, *sec*, *sed*, *see* and *ser* [40]. The second reaction enabled us to identify the *seg*,*seh*, *sei*, *sej* and *sep* genes [41].

4.5. Detection of Staphylococcal Enterotoxins A-E

The samples contaminated with CPS and all the final products were analyzed using a two-step method consisting of extraction/concentration and detection by enzyme linked fluorescent assay (ELFA) according to the European Union Reference Laboratory for Coagulase Positive Staphylococci protocol with the Vidas SET 2 test (bioMerieux) [42]. Detection of staphylococcal enterotoxins was also performed for *S. aureus* isolates harboring the sea-see genes after 72 h of incubation in brain heart infusion broth (BHI, Oxoid, Hampshire, UK) at 37 °C using the Ridasceen SET Total kit (R-Biopharm, Darmstadt, Germany).

4.6. Antimicrobial Resistance

A microbroth dilution method was used to establish the minimum inhibitory concentrations (MICs) of *S. aureus* isolates to 10 antimicrobial agents using the DKVP microplates (Trek Diagnostic Systems, Thermo Fisher Scientific, East Grinstead, UK). Antimicrobials, dilution ranges, and cut-off values used for MIC determination are described in Table 1. The isolates were cultured on Columbia agar supplemented with 5% sheep blood (bioMerieux) at 37 °C for 24 h. The MICs were established using Mueller-Hinton broth (Trek Diagnostic Systems, Thermo Fisher Scientific, East Grinstead, UK). The microplates were incubated at 36 °C for 18–20 h and read using the Vision® system (Swin Version 3.3, Trek Diagnostic System, Thermo Fisher Scientific, East Grinstead, UK, 2011). The cut off values for the interpretation of the MIC results were in accordance with the European Committee on Antimicrobial Susceptibility Testing (EUCAST) and the EURL-AR (Table 2).

Table 2. Antimicrobials, dilution ranges and cut-off values used for minimum inhibitory concentrations (MIC) determination of *S. aureus*.

Antimicrobial Class	Antimicrobials	Dilution Range (mg/L)	Cut off Values (mg/L) Resistant >
Amphenicols	Chloramphenicol (CHL)	2–64	16
Fluoroqinolones	Ciprofloxacin (CIP)	0.12–8	1
Macrolides	Erythromycin (ERY)	0.25–16	1
Aminoglycosides	Gentamicin (GEN)	0.25–16	2
	Streptomycin (STR)	4–64	16
β-Lactames	Penicillin (PEN)	0.06–16	0.125
Cephalosporins	Cefoxitin (FOX)	0.5–32	4
Tetracyclines	Tetracycline (TET)	0.5–32	1
Sulfonamides	Sulfamethoxazole (SMX)	32–512	128
Other	Trimethoprim (TMP)	0.5–32	4

AUTHOR CONTRIBUTIONS

J.G.R. conceived the study, designed the experiments, coordinated the investigation, analyzed the data and wrote the paper. A.C. and W.K-D. performed laboratory analysis and drafted the experimental section. J.O. critically reviewed the manuscript. All the authors have read and approved the final manuscript.

CONFLICTS OF INTEREST

The authors declare no conflict of interest.

REFERENCES

1. Argudin, M.A.; Mendoza, M.C.; Rodicio, M.R. Food poisoning and *Staphylococcus aureus* enterotoxins. *Toxins* **2010**, *2*, 1751–1773.

2. Hennekinne, J.A.; de Buyser, M.L.; Dragacci, S. *Staphylococcus aureus* and its food poisoning toxins: Characterization and outbreak investigation. *FEMS Microbiol. Rev.* **2012**, *36*, 815–836.

3. Le Loir, Y.; Baron, F.; Gautier, M. *Staphylococcus aureus* and food poisoning. *Genet. Mol. Res.* **2003**, *2*, 63–76.

4. Waage, S.; Mørk, T.; Røros, A.; Aasland, D.; Hunshamar, A.; Odegaard, S.A. Bacteria associated with clinical mastitis in dairy heifers. *J. Dairy Sci.* **1999**, *82*, 712–719.

5. Beuvier, E.; Buchin, S. Raw milk cheeses. In *Cheese: Chemistry, Physics and Microbiology*; Fox, P.F., McSweeney, P.L.H., Cogan, T.M., Guinee, T.P., Eds.; Elsevier Academic Press: Amsterdam, The Netherlands, 2004; pp. 319–345.

6. Williams, A.G.; Withers, S.E. Microbiological characterisation of artisanal farm house cheeses manufactured in Scotland. *Int. J. Dairy Technol.* **2010**, *63*, 356–369.

7. De Buyser, M.L.; Dufour, B.; Maire, M.; Lafarge, V. Implication of milk and milk products in food-borne diseases in France and in different industrialised countries. *Int. J. Food Microbiol.* **2001**, *67*, 1–17.

8. André, M.C.D.P.B.; Campos, M.R.H.; Borges, L.J.; Kipnis, A.; Pimenta, F.C.; Serafini, A.B. Comparison of*Staphylococcus aureus* isolates from food handlers, raw bovine milk and Minas Frescal cheese by antibiogram and pulsed-field gel electrophoresis following Smal digestion. *Food Control.* **2008**, *19*, 200–207.

9. Normanno, G.; Corrente, M.; La Salandra, G.; Dambrosio, A.; Quaglia, N.C.; Parisi, A.; Greco, G.; Ballacicco, A.L.; Virgilio, S.; Celano, G.V. Methicillin-resistant *Staphylococcus aureus* (MRSA) in foods of animal origin product in Italy.*Int. J. Food Microbiol.* **2007**, *117*, 219–222.

10. Loncarevic, S.; Jørgensen, H.J.; Løvseth, A.; Mathisen, T.; Rørvik, L.M. Diversity of *Staphylococcus aureus* enterotoxin types within single samples of raw milk and raw milk products. *J. Appl. Microbiol.* **2005**, *98*, 344–350.

11. Pinchuk, I.V.; Beswick, E.J.; Reyes, V.E. *Staphylococcal enterotoxins* . *Toxins* **2010**, *2*, 2177–2197.

12. EC—European Commission. Opinion on Staphylococcal Enterotoxins in milk products, particularly cheeses (adopted on 26–27 March 2003). 2003. Available online: http://ec.europa.eu/food/fs/sc/scv/out61_en.pdf (accessed on 26–27 March 2003).

13. EFSA and ECDC (European Food Safety Authority and European Centre for Disease Prevention and Control). The European Union summary report on trends and sources of zoonoses, zoonotic agents and food-borne outbreaks in 2013. *EFSA J.* **2015**, *13*, 3991.

14. Rola, J.G.; Korpysa-Dzirba, W.; Osek, J. Prevalence of *Staphylococcus aureus* and staphylococcal enterotoxins at different stages of production of raw milk cheeses—Preliminary results. *Bull. Vet. Inst. Pulawy.* **2013**, *57*, 341–345.

15. Bianchi, D.M.; Gallina, S.; Bellio, A.; Chiesa, F.; Civera, T.; Decastelli, L. Enterotoxin gene profiles of *Staphylococcus aureus* isolated from milk and dairy products in Italy. *Lett. Appl. Microbiol.* **2014**, *58*, 190–196.

16. Hill, B.; Smythe, B.; Lindsay, D.; Shepherd, J. Microbiology of raw milk in New Zealand. *Int. J. Food Microbiol.* **2012**,*157*, 305–308.

17. Hunt, K.; Schelin, J.; Radström, P.; Butler, F.; Jordan, K. Classical enterotoxins of coagulase-positive *Staphylococcus aureus* isolates from raw milk and products for raw milk cheese production in Ireland. *Dairy Sci. Technol.* **2012**, *92*, 487–499.

18. Jakobsen, R.A.; Heggebo, R.; Sunde, E.B.; Skjervheim, M. *Staphylococcus aureus* and *Listeria monocytogenes* in Norwegian raw milk cheese production. *Food Microbiol.* **2011**, *28*, 492–496.

19. Jorgensen, H.J.; Mork, T.; Hogasen, H.R.; Rorvik, L.M. Enterotoxigenic *Staphylococcus aureus* in bulk milk in Norway.*J. Appl. Microbiol.* **2005**, *99*, 158–166.

20. Peles, F.; Wagner, M.; Varga, L.; Hein, I.; Rieck, P.; Gutser, K.; Kereszturi, P.; Kardos, G.; Turcsanyi, I.; Beri, B.; *et al.* Characterization of *Staphylococcus aureus* strains isolated from bovine milk in Hungary. *Int. J. Food Microbiol.* **2007**, *118*, 186–193.

21. Tondo, E.C.; Guimaraes, M.C.; Henriques, J.A.; Ayub, M.A. Assessing and analysing contamination of a dairy products processing plant by *Staphylococcus aureus* using antibiotic resistance and PFGE. *Can. J. Microbiol.* **2000**, *46*, 1108–1114.

22. Callon, C.; Gilbert, F.B.; Cremoux, R.D.; Montel, M.C. Application of variable number of tandem repeat analysis to determine the origin of *S. aureus* contamination from milk to cheese in goat cheese farms. *Food Control.* **2008**, *19*, 143–150.

23. Asperger, H.; Zangerl, P. *Staphylococcus aureus* . In *Encyclopaedia of Dairy Sciences*; Roginski, H., Fuquay, J.W., Fox, P.F., Eds.; Academic Press: San Diego, CA, USA, 2003; pp. 2563–2569.

24. Tatini, S.R.; Jezeski, J.J.; Morris, H.A.; Olson, J.C., Jr.; Casman, E.P. Production of staphylococcal enterotoxin A in cheddar and Colby cheese. *J. Dairy Sci.* **1971**, *54*, 815–825.

25. Le Marc, Y.; Valík, L.; Medveďová, A. Modelling the effect of the starter culture on the growth of *Staphylococcus aureus*in milk. *Int. J. Food Microbiol.* **2009**, *129*, 306–311.

26. Lawrynowicz-Paciorek, M.; Kochman, M.; Piekarska, K.; Grochowska, A.; Windyga, B. The distribution of enterotoxin and enterotoxin-like genes in *Staphylococcus aureus* strains isolated from nasal carriers and food samples. *Int. J. Food Microbiol.* **2007**, *117*, 319–323.

27. Zhang, S.; Iandolo, J.J.; Stewart, G.C. The enterotoxin D plasmid of *Staphylococcus aureus* encodes a second enterotoxin determinant (*sej*). *FEMS Microbiol. Lett.* **1998**, *168*, 227–233.

28. Hummerjohann, J.; Naskova, J.; Baumgartner, A.; Graber, H.U. Enterotoxin-producing *Staphylococcus aureus* genotype B as a major contaminant in Swiss raw milk cheese. *J. Dairy Sci.* **2014**, *97*, 1305–1312.

29. Cremonesi, P.; Perez, G.; Pisoni, G.; Moroni, P.; Morandi, S.; Luzzana, M.; Brasca, M.; Castiglioni, B. Detection of enterotoxigenic *Staphylococcus aureus* isolates in raw milk cheese. *Lett. Appl. Microbiol.* **2007**, *45*, 586–591.

30. Morandi, S.; Brasca, M.; Lodi, R.; Cremonesi, P.; Castiglioni, B. Detection of classical enterotoxins and identification of enterotoxin genes in *Staphylococcus aureus* from milk and dairy products. *Vet. Microbiol.* **2007**, *124*, 66–72.

31. Katsuda, K.; Hata, E.; Kobayashi, H.; Kohmoto, M.; Kawashima, K.; Tsunemitsu, H.; Eguchi, M. Molecular typing of *Staphylococcus*

aureus isolated from bovine mastitic milk on the basis of toxin genes and coagulase gene polymorphisms. *Vet. Microbiol.* **2005**, *105*, 301–305.

32. Villard, L.; Lamprell, H.; Borges, E.; Maurin, F.; Noel, Y.; Beuvier, E.; Chamba, J.F.; Kodjo, A. Enterotoxin D producing strains of *Staphylococcus aureus* are typeable by pulsed-field gel electrophoresis (PFGE). *Food Microbiol.* **2005**, *22*, 261–265.

33. Kav, K.; Col, R.; Ardic, M. Characterization of *Staphylococcus aureus* isolates from white-brined Urfa cheese. *J. Food Prot.* **2011**, *74*, 1788–1796.

34. Werckenthin, C.; Cardoso, M.; Martel, J.L.; Schwarz, S. Antimicrobial resistance in staphylococci from animals with particular reference to bovine *Staphylococcus aureus* and porcine *Staphylococcus hyicus* and canine *Staphylococcus intermedius*. *Vet. Res.* **2001**, *32*, 341–362.

35. Krasucka, D.; Cybulski, W.; Klimowicz, A.; Dzierżawski, A. Evaluation of antimicrobial agents consumption in swine and cattle in Poland based on a questionnaire in 2010. *Medycyna. Wet.* **2012**, *68*, 102–105.

36. European Medicines Agency, European Surveillance of Veterinary Antimicrobial Consumption. Sales of veterinary antimicrobial agents in 26 EU/EEA countries in 2012. (EMA/333921/2014). Available online: http://www.ema.europa.eu/docs/en_GB/document_library/Report/2014/10/WC500175671.pdf (accessed on 15 October 2014).

37. Berendsen, B.; Stolker, L.; de Jong, J.; Nielsen, M.; Ttserendorj, E.; Sodnomdarjaa, R.; Cannavan, A.; Elliott, C. Evidence of natural occurrence of the banned antibiotic chloramfenicol in herbs and grass. *Anal. Bioanal. Chem.* **2010**,*397*, 1955–1963.

38. ISO 6888–2, 1999. In *Microbiology of Food Animal Feeding Stuffs—Horizontal Method for the Enumeration of Coagulase-Positive Staphylococci (Staphylococcus Aureus and Other Species). Part 2: Technique Using Rabbit Plasma Fibrynogen Agar Medium*; International Standard Organisation: Geneva, Switzerland, 1999; pp. 1–7.

39. Anonymous. Multiplex PCR for the detection of *mec*A gene and identification of *Staphylococcus aureus*. Available online: http// www.crl-ar.eu/data/images/tc_april-2009/6-detailed%20meca-pcr_protocol.pdf (accessed on 20 April 2009).

40. De Buyser, M.L.; Grout, J.; Brisabois, A.; Assere, A.; Lombard, B. *Detection of Genes Encoding Staphylococcal Enterotoxins. Multiplex PCR for Sea to See and Ser. Method of the CRL for Coagulase Positive Staphylococci Including Staphylococcus aureus*, 1st ed.; CRL CPS; AFSSA: Maisons-Alfort, France, 2009; pp. 1–5.

41. De Buyser, M.L.; Grout, J.; Brisabois, A.; Assere, A.; Lombard, B. *Detection of Genes Encoding Staphylococcal Enterotoxins. Multiplex PCR for Seg to Sej and Sep. Method of the CRL for Coagulase Positive Staphylococci Including Staphylococcus aureus*, 1st ed.; CRL CPS; AFSSA: Maisons-Alfort, France, 2009; pp. 1–5.

42. Ostyn, A.; Prufer, A.L.; Papinaud, J.; Hennekinne, J.-A.; Assere, A.; Lombard, B. *Detection of Staphylococcal Enteritoxins Types SEA to SEE in All Types of Food Matrices. European Screening Method of the EU-RL for "Coagulase Positive Staphylococci Including Staphylococcus aureus"*, 5th ed.; EU-RL CPS; ANSES: Maisons-Alfort, France, 2010; pp. 1–12.

CHAPTER 7

Digital Dermatitis in Dairy Cows: A Review of Risk Factors and Potential Sources of Between-Animal Variation in Susceptibility

Maeve A. Palmer * and Niamh E. O'Connell

Institute for Global Food Security, School of Biological Sciences, Queen's University Belfast, 18-30 Malone Road, Belfast,

Northern Ireland, BT9 5BN, UK

SIMPLE SUMMARY

Dairy cow lameness is a major problem for the industry, causing reduced animal welfare and economic loss. Digital dermatitis (DD) is a bacterial disease causing painful lesions, generally on the heels of the rear feet, and is an important cause of lameness. There appears to be individual variation between animals in susceptibility to this disease. Particular physical, physiological and behavioural factors might influence individual susceptibility, but further work is required to clarify the influence of these factors and to determine how this information could be used to develop breeding and management strategies to reduce DD prevalence.

ABSTRACT

Digital dermatitis (DD) is a bacterial disease that primarily affects the skin on the heels of cattle. It is a major cause of lameness in dairy cows and a significant problem for the dairy industry in many countries, causing reduced animal welfare and economic loss. A wide range of infection levels has been found on infected farms, prompting investigations into both farm level and animal level risk factors for DD occurrence. There also appears to be individual variation between animals in susceptibility to the disease. The identification of factors affecting individual variation in susceptibility to DD might allow changes in breeding policies or herd management which could be used to reduce DD prevalence. Factors mentioned in the literature as possibly influencing individual variation in susceptibility to DD include physical factors such as hoof conformation and properties of the skin, physiological factors such as the efficacy of the immune response, and behavioural factors such as standing half in cubicles. Further work is required to determine the influence of these factors, identify the genetic basis of variation, clarify the level of heritability of DD susceptibility and to determine how this is correlated with production and health traits currently used in breeding programmes.

KEYWORDS

digital dermatitis; dairy cow; lameness; susceptibility

1. INTRODUCTION

Digital dermatitis (DD) is a bacterial disease that primarily affects the skin on the heels of cattle. Infection causes inflammation and skin damage, leading to pain and discomfort [1]. It is a major cause of lameness in dairy cows [2] and hence a significant problem for the dairy industry in many countries, causing reduced animal welfare and economic loss [3]. DD has recently been identified as an emerging issue in beef cattle in the UK [4] and the bacteria believed to cause DD have been identified in similar lesions in sheep (known as Contagious Ovine Digital Dermatitis; [5]), dairy goats [6,7] and even wild North American Elk [7]. In addition to this, these DD associated bacteria have been detected in three types of severe bovine foot lesions which have emerged during the last 15 years; toe necrosis, non-healing white line disease and non-healing sole ulcer [8]. These developments highlight the growing importance of DD for domestic and wild animals, and for farmers and veterinarians.

Despite the economic and welfare importance of this disease, many questions remain regarding its etiology, transmission, prevention and treatment. There are a number of reasons why the disease is proving difficult to deal with; firstly, the infection appears to be polymicrobial, with a variety of bacteria, particularly of the genus *Treponema*, isolated from lesions [9,10,11,12,13,14,15]. In addition, the bacteria involved initially proved difficult to grow in culture [13], experimental infection models have been difficult to develop [16,17], and the mechanisms of disease transmission have thus remained rather mysterious. Recent advances in laboratory methods have meant that good progress has been made in the identification of the most important pathogenic bacteria, and in the detection of these bacteria in the animals and in the environment of farms with DD infections [14,15,18,19].

A wide range of infection levels has been found on infected farms, prompting investigations into both farm/herd level risk factors and animal level risk factors (for example parity and stage of lactation) for DD occurrence (e.g., [20,21] summarised by [22]). Both the farm level and animal level risk factors can provide useful information when trying to minimise DD infection levels and understand when risks of infection are highest. An interesting, but less investigated, aspect of DD is that there appears to be individual variation between animals in susceptibility to the disease; Laven [23,24], Capion *et al.* [24] and Gomez *et al.* [25] all found that some animals within a herd were infected repeatedly while others of the same breed and parity and kept under the same conditions were never infected. If the reasons for this individual variation in susceptibility could be identified, they could add to our understanding of the disease and contribute to the search for effective prevention and treatment methods.

This review will begin by providing an overview of the disease and the problems that it causes, focusing where possible on new developments reported in the literature. It will then briefly outline the farm-level and animal-level risk factors for DD that have been described in the literature. Finally, it will introduce research that has been carried out into individual variation in susceptibility to DD.

2. BACKGROUND

DD was first described in Italy in 1974 [26] but has since been recorded in the majority of countries with a dairy industry [27]. Infection causes ulcerative lesions along the coronary band [26] or on the skin adjacent to the interdigital cleft [28], leading to pain and discomfort [1]. The course of the disease is generally described as according to Döpfer *et al.* [29], who split it into four stages. The early lesion or M1 stage is a circumscribed granulomatous area that is normally small in size [29] and is not generally painful [28]. The lesion then develops into the M2 or classic ulcerative stage [29]. This stage of the lesion is generally larger and is painful on palpation [28]. Once the lesion begins to heal (which is not normally until after treatment), a scab forms over the ulcerated area and the lesion is described as an M3 [29]. In some cases the lesion progresses to an M4 or chronic stage [29] which is characterised by surface proliferation or dyskeratosis. This stage is generally not painful but is infectious and can transition back to an active M1 lesion [30,31]. The M4.1 lesion type was added by Berry *et al.* [31] to describe a chronic lesion (M4) with a small, painful area of active M1 lesion within it. Once a lesion has completely healed and only healthy skin remains it is classified as an M5 [31]. It is common to find animals with lesions on both rear feet concurrently; Laven [23] found that 51% of animals in one herd had lesions on both feet, 22% on the left foot only and 27% on the right foot only. It is not known why some animals develop DD on only one hind foot when both feet are exposed to the same risk factors, and investigation of the factors influencing which foot becomes infected (or which is infected first) could contribute to the search for factors which affect variation in susceptibility to the disease.

The presentation of DD varies geographically, with a proliferative form more common in the USA and a more erosive form predominant in Europe [32]. The disease is generally referred to as papillomatous DD (PDD) in the USA and digital dermatitis (DD) in Europe [32], however during this review the term digital dermatitis (DD) will be used to refer to both conditions.

2.1. Etiology and Transmission

Although a lot of questions still remain, recent advances in laboratory techniques such as gene sequencing have begun to provide the answers to essential questions regarding the identification of causative bacteria and the transmission of the disease. Spirochaete bacteria of the genus *Treponema* are consistently found in DD lesions (e.g., [9,10,12,14,15,33,34]) and are thought likely to be the primary aetiological agents of DD [35]. Their presence deep within lesions [36] suggests that they are invasive and not simply colonising damaged tissue [37]. Treponemes have been isolated from tissue-destructive disease lesions from a number of different hosts, including humans [37].

Treponemal diseases in humans include venereal and endemic syphilis, bejel, yaws and pinta, which all have a relapsing clinical course and include (but most are not restricted to) skin lesions [38]. Treponemes have also been implicated in human periodontitis, which is also thought to have a polymicrobial etiology [37], and indeed some of the treponemes identified from DD lesions show strong similarities to those which are involved in periodontitis [10]. Although treponemes are the most commonly found bacteria in lesions, other types of bacteria have also been identified including *Borrelia burgdorferi* [36,39], *Bacteroides* and *Mycoplasma* species [39],*Campylobacter* species [29] and *Candidatus Amoebophilus asiaticus* [15].

Recently, work has been carried out to characterise the microbiome of healthy skin and DD lesions at different stages of infection [14,15]. Krull *et al.* [14] examined the microbiota of lesions at seven different stages of development and found that there was a dramatic increase in the proportion of treponemes detected in lesion biopsies as the lesions developed. They also found that the proportions of the different treponeme species identified changed as the lesions developed. This agrees with the findings of Zinicola *et al.* [15], who found that the microbiome of skin with active DD lesions (M1, M2 and M4.1) was distinct from the microbiome of skin with inactive lesions (M3 and M4) and from healthy skin. The microbiome of active, ulcerative DD lesions was dominated by six groups of treponemes; *Treponema denticola*, *Treponema maltophilum*,*Treponema medium*, *Treponema putidum*, *Treponema phagedenis* and *Treponema paraluiscuniculi* [15]. The findings of these two studies suggest that some of the variation in bacterial species found in DD lesions between previous studies could be due to samples being taken at different stages of lesion development.

The advances in laboratory techniques which are helping to clarify the causative agents of DD are also helping to show how the disease might spread between animals. Although environmental transmission of DD-causing treponemes has been postulated for many years, these bacteria initially proved hard to find in the environment, with Evans *et al.* [18] not finding any evidence of them in dairy cow faeces or environmental slurry samples. Since then, advances in technology have allowed the detection of treponemes in both dairy cow faeces and environmental slurry samples on infected farms by using a targeted deep-sequencing approach to detect very small amounts of bacterial RNA [19]. The role of the gut as a possible reservoir of DD bacteria was postulated by Shibahara *et al.* [40], who found that similar spirochaetes appeared to be responsible for simultaneous infections of bovine dysentery and DD in two Japanese dairy cows. Since then, Evans *et al.* [18] detected DD-related treponemes in both oral and rectal tissues of dairy cows on DD-affected farms, and these findings have been confirmed by Zinicola *et al.* [15], who found DD-associated treponemes in both the rumen and faecal microbiomes.

These findings suggest that the gastrointestinal tract acts as a reservoir for DD-related treponemes [15] and faeces/slurry as a mode of transmission between the reservoir and site of infection [18]. Direct skin to skin transmission from infected to uninfected feet has also been suggested as a possible route of infection [18], as has transmission via hoof trimming implements [41].

2.2. Treatment and Prospects for Vaccine Development

Treatments for DD include systemic and topical antibiotics [1,9] and are summarised by [22]. In herds where a high proportion of animals are infected with DD individual treatment is very time-consuming, so many farmers instead use footbaths to treat the entire herd [1]. Unfortunately, elimination of DD is rarely seen, so repeated application of treatments is required to prevent recurrence of infection [2]. An antibody response is produced by dairy cows when they are infected with DD [33,42], however this response does not seem sufficient to prevent further infections, as some animals are infected repeatedly [42]. One reason suggested for the difficulties encountered when trying to treat DD and prevent recurrence is the fact that a number of the treponemes thought to be involved in DD have been shown to have encysted as well as spiral forms [43]. It is possible that these encysted forms of the bacteria could persist deep within the lesions and cause a recurrence of clinical disease at a later date, though more research is required to determine the significance of the encysted form of the bacteria and its response to DD treatments [43].

There have been a number of efforts over at least 15 years to develop a vaccine to DD, however none are currently available. Two studies have found promising results. In the first, Keil *et al.* [44] found that adult cows and heifers vaccinated with an inactivated *Treponema* bacterin showed a significant reduction in PDD prevalence compared to unvaccinated control animals. In the second, Berry *et al.* [45] report a study in Nebraska where a different *Treponema* bacterin marketed as TrepShield (Novartis Animal Health) led to a significantly lower occurrence of DD in vaccinated animals when heifers were immunised before calving and cows immunised during the dry period, however this vaccine has since been withdrawn.

The results of other trials have been less encouraging. Berry *et al.* [46,47] vaccinated and then experimentally infected calves with four species of bacteria and found mixed results—the diameter of lesions was reduced in vaccinated calves in one experiment but not in another and no antibody response to the vaccine developed in either experiment. Fiddler *et al.*[47]

report the findings of a study which evaluated a vaccine containing *Serpens* sp. bacterin, developed by a commercial company. They found that, although the dairy cows involved in the trial did develop an immune response to the bacterin, there was no reduction in the prevalence or severity of DD infections among vaccinated cows when compared to controls. They therefore concluded that the vaccine did not show any clinical efficacy in terms of DD prevention [47]. Work is ongoing to develop a vaccine which will be effective in the prevention of DD. For example, Staton *et al.* [48] are currently using a reverse vaccinology approach to identify vaccine candidate proteins for a recombinant vaccine which they hope will be effective against treponemes from three different phylogenetic groups.

2.3. Consequences of DD

The majority of active DD lesions are painful [1,21,49], so the primary consequence of DD infection is likely to be pain. If the DD is not treated, infected animals can remain lame for up to four months [50], implying ongoing pain and discomfort. Animals that are in sufficient pain to develop lameness also show other changes in behaviour when compared to sound animals, such as an increase in lying time [51,52,53,54] and shorter total feeding time [51,53,55,56], which in turn might be expected to have some effect on productivity.

Studies that have examined the effect of DD on milk yield have given differing results and, although it is a major cause of lameness, DD infection does not always appear to be associated with a significant reduction in milk yield. Amory *et al.* [57] found that animals with DD in England and Wales did not show a decreased milk yield during infection, but had a slightly increased milk yield after treatment. In two other studies, animals with DD produced less milk, but the difference was not significant [58,59]. Warnick *et al.* [60] found that animals with DD on two US dairy farms showed a reduction in milk yield, but found that this reduction was not as large as for animals with some other causes of lameness. Using a large data set from Holstein cows on French farms, Relun *et al.* [61] found that DD caused a small but significant decrease in milk yield (<1 kg per day). Pavlenko *et al.* [62] found that DD-affected Swedish Red and Swedish Holstein cows had a significantly lower milk yield (5.5 kg energy corrected milk per day) than healthy control cows. Gomez *et al.* [25] found that the DD infection history of heifers affected their milk yield during the first lactation; a reduction in milk yield of 199 kg or 335 kg milk over 305 days was found for cows that had experienced one or more than one DD infection before their first calving respectively, when compared to cows that had not experienced a DD infection before calving.

Although there is evidence that lameness in general impairs fertility (e.g., [63,64,65]), there has been very little work relating DD infection specifically to changes in fertility. An increase in the calving to conception interval and the number of days open was seen in Mexican Holstein-Friesian cows with DD compared to healthy cows [58]. In agreement with this, Gomez *et al.* [25] found that heifers repeatedly infected by DD before their first calving had a lower conception rate at first service and an increase in the number of days open when compared to cows that had not had any DD infection before first calving. Overall, Zinicola *et al.* [15] estimate that DD leads to an annual economic loss of US $1.1billion from dairy cows in the US and EU if a DD prevalence rate of 25% is assumed.

3. FARM LEVEL FACTORS AFFECTING SUSCEPTIBILITY TO DD

A number of studies conducted across different countries have examined farm level risk factors for DD (e.g., [21,66,67,68,69]). The main risk factors found relating to housing and general management are briefly discussed below.

3.1. Housing

The majority of the research that has examined the relationship between pasture or confinement and DD has found that increased access to pasture is associated with a decreased risk of DD [21,32,68,70]. In contrast to this, one study found that access to pasture was associated with an increased risk of DD [69]. However the authors suggest that this relationship may in fact be an effect of DD; caused by herdsmen providing additional access to pasture when it is considered that the level of claw disorders is high.

The type of housing provided for dairy cows has also been linked to DD prevalence. Laven [23] showed that cattle housed in straw yards had less, and less severe, DD than those housed in cubicles. Onyiro *et al.* [70] also found that cubicle housed animals were more likely to suffer from DD than animals in straw yards or at pasture. Within cubicle houses, longer and wider cubicles were found to reduce the risk of the disease [21]. A suggested connection between cubicle size and risk is that if cubicles are too small, animals are likely to spend more time standing, which increases the contact between their heels and slurry [71].

A number of authors have found links between the cleanliness of the environment and DD prevalence. In fact, Relun *et al.*[72] found a direct relationship between levels of leg cleanliness in dairy cows (measured at the herd level) and levels of DD; herds in which animals had dirtier legs had a higher prevalence of DD. Solid floors composed of grooved concrete were identified as a risk in the UK [73] and in the USA [68] when compared to textured concrete floors. This could be due to the fact that a small amount of slurry remains in the grooves of this type of flooring even when it is freshly scraped [73]. A common theme throughout the housing factors that have been linked to DD is the level of contact between the cows' feet and slurry, which has been highlighted as an important risk factor for DD [23]. In addition, the moisture level in the environment (which could be affected by both the efficiency of slurry removal and building ventilation) could also be important as DD has been associated with damp conditions both in cubicle houses [20] and in corrals [32].

3.2. Biosecurity—External Hoof Trimmers and Buying in Replacements

The introduction of new animals to the herd, primarily through buying in replacement heifers, has been found to be associated with DD prevalence by studies in the USA and Chile [66,67,68]. Certain hoof trimming practices have also been highlighted as a biosecurity concern and have been linked to DD prevalence. These include the use of professional hoof trimmers (who work on a number of different farms), and hoof trimmers who do not clean the hoof trimming equipment between animals [68]. Evidence in support of the theory that the bacteria causing DD could be spread by hoof trimming equipment was recently provided by Sullivan *et al.* [41], who found that treponemes from at least one of the three phylogroups associated with DD (*T. medium*/*T. vincentii*-like, *T. phagedenis*-like and *T. denticola*/*T. putidum*-like) were present in at least 42% of swabs from knives used to trim the hooves of cattle infected with DD. Disinfection of the knife with an iodine disinfectant greatly reduced the proportion of swabs that tested positive for the treponemes, but did not eliminate the treponemes entirely (with 24% of swabs from disinfected knives testing positive for at least one phylogroup). A number of papers have, however, found that hoof trimming was associated with a reduced risk of DD occurrence [21,69,72]. Relun *et al.* [72] suggest that this could be because the DD lesions were treated at hoof trimming visits and then recovered quickly, or because hoof trimming could have improved the conformation of the feet, making the animals less susceptible to DD (as discussed in Section 5.3).

3.3. Diet and Nutrition

Feeding strategy and nutritional content of the diet have both been identified as possible risk factors for DD. The rate at which concentrate supplementation is increased for dairy cattle after parturition has been linked to DD prevalence; Somers*et al.* [21] found a higher incidence of DD in herds where the maximum level of supplementation was reached at two weeks post-partum (compared to at three weeks). The authors suggested that this could be due to an increased metabolic imbalance leading to a greater susceptibility to disease.

Gomez *et al.* [74] recently investigated whether the beneficial effects of dietary supplements on immune responses and skin quality might reduce the incidence of DD. They supplemented Holstein steers with higher than standard levels of organic trace minerals and iodine and then monitored their responses to natural DD infection and experimental infection. No natural DD infections occurred during the trial, and the infection rates following experimental infection were lower than predicted. Therefore, no significant differences were found between the control and treatment groups, however there was a trend for the supplemented animals to have a reduced DD infection rate and also a trend for the average size of the experimentally induced lesions to be smaller in this group [74].

3.4. Management Practices

A number of general farm characteristics and management practices have been associated with the risk of DD. Two studies have found that the risk of DD increases with increasing herd size [66,68]. The way in which animals are grouped has also been found to have an effect; Somers *et al.* [21] found that the incidence of DD was reduced if heifers were housed with the lactating dairy cows for a prolonged time before calving, which the authors suggested gave the heifers more time to adapt to the environment. The same study also found that housing dry cows with the milking herd before calving actually increased their risk of DD; risk was lowest if they were introduced to the milking herd immediately after calving [21].

4. ANIMAL LEVEL FACTORS AFFECTING SUSCEPTIBILITY TO DD

A number of risk factors have also been reported which identify groups of animals within a herd that are at a higher risk for DD than others; animals of particular parity, animals at particular stages of lactation and animals of a particular breed. These factors are likely to affect all animals within the group in a similar way, however there is some variation in the literature in terms of which parity of animals and stage of lactation are associated with highest risk of DD infection, as discussed below.

4.1. Parity

A number of studies have found that first parity animals are most at risk of developing DD [21,32,67]. It has been suggested that the increased prevalence in this group is caused by the stress related to the significant environmental and metabolic changes that heifers experience around the time of their first calving [21,32,67]. In contrast to this, two studies have found that second lactation animals are the most likely to suffer from DD [58,69]. There may be management differences affecting which parity is most at risk. For example some farms may keep heifers separate from the main dairy herd for the first lactation, meaning that they do not experience the stress of mixing with the older cows until the beginning of the second lactation. Somers *et al.* [21] and Holzhauer *et al.* [69] both found that the risk of DD declined with each parity after the second lactation. The fact that there is such a high recurrence rate of DD lesions suggests that the lower incidence in later lactations is not due to the development of immunity to the disease [32]. A contributory factor to the decline in lesion prevalence with age may be that older animals with lesions are culled [67]. If DD susceptibility is linked to hoof conformation (see Section 5.3) then there may be a further reason for prevalence declining with parity as heel height increases as animals age [75].

4.2. Stage of Lactation

It is generally agreed that cows in the dry period have a lower risk of developing DD than lactating cows [21,28,58,69]. Dry cows are generally fed a different diet which may lead to more solid faeces, reducing the contact between the cows' heels and liquid slurry [21]. There is not yet a clear answer, however, to the question of which period during lactation animals are most susceptible to DD. Argaez-Rodriguez *et al.* [58] found that the highest risk for new infections was in the first month after calving. There is immune suppression in the peri-partum period making animals more susceptible to disease [76], and this may be enhanced in cows experiencing metabolic stress [77]. These factors could, in turn, contribute to higher levels of DD [21]. However, Holzhauer *et al.* [69] found that animals were most at risk during peak lactation. The high energy diet fed at peak lactation has been suggested as a contributing factor to the high prevalence of DD lesions at this time—a high energy diet can lead to larger quantities of a more liquid faeces being produced and hence more contamination of the heels with slurry [21,69,71].

4.3. Breed

A number of studies have found differences between breeds in lameness and DD prevalence, and many of these have found that Holstein-Friesians (and their cross breeds) are more susceptible than other breeds. For example, Barker *et al.* [73] found that farms with no Holstein-Friesian cows had a lower prevalence of lameness than farms with only Holstein-Friesians. In agreement with this, Holzhauer *et al.* [69] found that purebred and crossbred Holstein-Friesian cows had a higher risk of DD than a dual purpose breed (MRIJ), Rodriguez-Lainz *et al.* [67] found that Holstein x German Black-Pied animals had a higher prevalence of PDD than German Red-Pied animals in Chilean dairies, and Relun *et al.* [72] found that Holsteins were more likely to develop DD than Normande or other breeds on a group of French dairy farms.

5. INDIVIDUAL FACTORS AFFECTING SUSCEPTIBILITY TO DD

Individual differences in susceptibility to a number of infectious diseases have been investigated in dairy cows; for example mastitis [78,79], uterine diseases [80] and Johne's disease [81,82]. Variation in susceptibility to mastitis has been particularly well researched, and a number of factors which influence individual susceptibility have been identified [78]. These factors are both morphological, for example udder shape and placement [83], and physiological, such as aspects of immune function (e.g., [84,79]. It could be expected that individual variation in susceptibility to DD would also be affected by morphological and physiological factors. As it seems likely that the transmission of DD involves reservoirs of infection in the environment [15,19] and might also occur by direct cow-to-cow transmission [18], it seems pertinent to also consider the behaviour of the animals as a source of individual variation in susceptibility, as this could affect their exposure to environmental reservoirs and infected animals.

The majority of morphological, physiological and even behavioural factors which could influence individual susceptibility to DD are likely to be influenced to a greater or lesser extent by genetics. This means that there are two possible strategies which can be used when trying to identify individual differences which affect susceptibility; searching for the underlying genetic differences or examining the morphological, physiological or behavioural changes that these genetic differences produce or contribute to (phenotypic differences). The majority of work carried out to date into DD susceptibility has focused on phenotypic differences, however the rapid advancement of genomics makes it a promising area for future research, allowing precise determination of where genetic differences lie. This section will begin by outlining research that has been carried out into the heritability of DD and the use of genetic approaches in the investigation of individual differences in susceptibility. It will then go on to outline the aspects of variation in some physical (hoof conformation, properties of the skin), physiological (immune response) and behavioural characteristics which could contribute to individual differences in susceptibility to DD.

5.1. Heritability of DD

Genetic heritability values found for DD in Holstein-Friesian dairy cows differ quite widely between studies. van der Waaij *et al.* [85] calculated a heritability of 0.10 for DD in Dutch Holstein-Friesian dairy cows, considerably higher than the value of 0.029 calculated by Onyiro *et al.* [70] for DD in British Holstein-Friesian dairy cows. It is suggested by Onyiro *et al.*[70] that this difference could be due to the difference in models used to estimate heritability or to the size of the dataset used (over 90,000 records used by Onyiro *et al.* [70] compared to approximately 22,000 used by van der Waaij *et al.* [85]). A substantially higher heritability of between 0.33 and 0.40 for DD was found by Oberbauer *et al.* [86], who used repeated observations on 5043 Holstein-Friesian cows over a number of lactations on three dairy farms in California, USA. The authors suggest that the similar production systems used on the three farms (freestall housing) and the consistent classification of lesions by trained foot trimmers could have contributed to the higher heritability that they found [86].

Although there is variation between studies in the estimated heritability of DD, the fact that all of the studies have found a heritable component demonstrates that there is scope to selectively breed for a reduction in DD susceptibility. The consequences of selection for a reduction in DD are, however, not clear as these is no agreement as to the genetic correlations between DD and milk yield. Onyiro *et al.* [70] found that presence of DD was associated with lower lifespan and decreased milk and fat yield, so suggested that breeding for reduced DD could be accompanied by an improvement in these traits. More recently, however, Oikonomou *et al.* [87] found that cows whose sires had high genetic merit for milk and protein production were at higher risk of developing DD.

5.2. Genetic Approaches to Investigating DD Susceptibility

Scholey *et al.* [88] investigated variation in individual susceptibility to DD from a genetic perspective, by looking for single nucleotide polymorphisms (SNPs) which differed between animals that had a history of DD and those that appeared unaffected by DD. The study found eight SNPs that were associated with DD disease status, some of which were located in genes which could have a plausible role in DD pathology, such as skin cell proliferation and inflammatory responses [88]. It has been suggested that this type of study could be repeated with larger sample sizes as the costs of genotyping decrease [88] and that the information gained from such work could be used in screening programmes as part of targeted breeding strategies to reduce the number of animals susceptible to DD [89].

5.3. Hoof Conformation

One of the first physical attributes investigated with respect to variation in DD susceptibility was the conformation of cows' hooves. There is considerable variation in hoof conformation within breeds of dairy cattle and at least a proportion of this variation is heritable [90]. As hoof conformation could affect the amount of contact between heels and slurry or damp underfoot conditions (shown to be important for the establishment of DD infection by [16,17], a number of studies have been carried out to examine relationships between hoof shape and DD occurrence. Although a pilot study by Walker *et al.*[91] did not find any relationship between DD presence and claw length, claw width or claw angle (the slope of the dorsal border of the claw with respect to the floor surface [75]), a number of studies since then have found relationships between aspects of hoof conformation and DD.

Laven [92] carried out a preliminary study of DD and hoof conformation in 20 Holstein heifers, monitoring their DD status from three months of age until 12 weeks after calving and making hoof conformation measurements 24 weeks after calving. He found that at this stage the mean heel height was lower and mean toe length was longer for animals that had DD than for animals that did not [92]. In agreement with this, lower heel height associated with a higher prevalence of DD was also found by Olechnowicz and Jaśkowski [93] in a study of 144 primiparous Polish Holstein-Friesian cows. Laven [92] suggested that lower heels could lead to a higher risk of DD by increasing the amount of contact between the skin on the heels and slurry. Not all findings of the two studies agreed, however, as Olechnowicz and Jaśkowski [93] also found that a higher risk of DD was associated with a shorter dorsal length (distance between the dorsal border and the tip of the toe).

Baird *et al.* [94] examined a total of 84 Holstein-Friesian and Norwegian dairy cows in three production systems (grass based; housed and fed a low concentrate diet; housed and fed a high concentrate diet) in Northern Ireland. They found that animals with DD had wider heels than animals without DD, but did not find any relationship between heel height or claw length and DD occurrence. There has also been some interest in the role that could be played by the interdigital cleft; Daniel [95] found a pattern of increasing DD prevalence with decreasing width of the interdigital cleft in Holstein cows and postulated a link between DD development and the more anaerobic conditions that a narrow interdigital space would produce.

For the purpose of this review, one drawback of the studies discussed above [92,93,94,95] is their cross-sectional design, which means that it was not possible for them to demonstrate that any differences observed were likely to be the cause (rather than an effect) of DD [92,96]. A longitudinal study of the effect of DD on hoof conformation has recently been carried out by Gomez *et al.* [96] using a cohort of 644 Holstein heifers. Monitoring the animals for DD lesions and taking three sets of hoof measurements over approximately six months (starting at a mean age of 540d), they found that M2 (active) DD lesions were associated with increased heel height, claw angle and depth of the interdigital cleft (the distance between the skin-horn junction of the lateral and medial heels and the proximal end of the skin fold at the plantar aspect of the foot [96]). The increased heel height found in this study is attributed to the fact that cows with DD tend to take weight off their heels and onto their claws, which increases wear of the toes and reduces that of the heel, thus increasing heel height over time [92,96]. The same process is likely to result in the increased claw angle observed [96]. A greater claw angle was found in animals with DD than in animals without DD by Olechnowicz and Jaśkowski [93] and the findings of the Gomez *et al.* [96] study suggest that this could be due to DD rather than a predisposing factor. The increase in the depth of the interdigital cleft associated with active DD lesions was thought by Gomez *et al.* [96] to be caused by inflammation and the prominence of the DD lesion. These changes caused by the active lesion could cause a more anaerobic environment and potentially facilitate further bacterial growth [96].

It is clear that studies investigating the relationships between hoof conformation and susceptibility to DD have not found consistent results and there are a number of factors which could contribute to this, including the small sample sizes used in many of the studies, the varied housing and management systems and the complicating point that DD has now been shown to alter hoof conformation as lesions progress [96]. Further, larger-scale research in this area could help to clarify whether some aspects of hoof conformation, including heel height and the morphology of the interdigital cleft, are associated with individual susceptibility and whether relationships are similar in different breeds and housing systems.

5.4. Properties of the Skin and Hair Follicles

A second physical characteristic which has been proposed as a possible cause of differences in susceptibility to DD is the efficacy of the skin in preventing pathogen entry [97]. How well the skin acts as a barrier to particles and pathogens can vary between individuals and is affected by a number of factors including the thickness [98] and lipid content [99] of the stratum corneum layer (the top layer of the epidermis). The barrier function can also be affected by single gene mutations with large effects; for example some mutations in the gene that codes for the protein filaggrin (important in correct stratum corneum formation) have been shown to reduce barrier function and are associated with the occurrence of atopic dermatitis in humans [100].

In a small study comparing the skin permeability of dairy cows with and without a history of DD, Palmer *et al.* [97] did not find any difference in the permeability of skin to a dye solution (methylene blue). The study also tested the effect of prolonged (24 hour) contact with slurry on the permeability of the skin and found that slurry increased skin permeability to the dye, but that there was no difference in the magnitude of the increase in permeability between animals with and without a history of DD [97]. These results imply that in the population investigated, differences in susceptibility to DD were unlikely to be caused by simple differences in the skin barrier function (as is the case in some humans with atopic dermatitis and accompanying skin infections [101]).

There has been growing interest in recent years in the role that the hair follicles might play in the movement of particles across the skin barrier [102,103,104]. The stratum corneum is not entirely continuous around the base of hair follicles which means that there are areas where it is easier for particles to pass through into the lower layers of the skin [104]. In relation to DD, it has been suggested that treponemes might use the hair follicles as a means of entry to, or exit from, bovine skin, as electron micrographs of DD-infected tissues have shown large numbers of treponemes in the hair follicles and radiating out towards deeper parts of the tissues [34]. The possible role of hair follicles in the development of DD lesions has been further highlighted in a study by Scholey *et al.* [105] which examined changes in gene expression within host tissues during DD infection. They found that there was altered expression of a number of genes relating to epidermal barrier formation within DD lesions and they suggested that this could reduce the efficacy of the barrier between the hair shafts and skin [105].

Hair follicles can also act as a reservoir for bacteria and therefore a source for re-colonisation after treatment and disinfection [106]. If this is the case for DD it would have implications for treatment, as it would highlight the importance of getting any treatment solutions deep into the hair follicles. It might also be possible to develop treatments aimed specifically at the hair follicles which could then act as reservoirs of antibacterial activity, as has been investigated for *Candida albicans*infection in human skin [107]. Considerable variation has been found between individuals in the number of hairs per unit area in Brazilian Holstein cattle (a range of 221 to 3545 hairs per cm^2, [108]). A proportion of the variation in number of hairs per unit area does appear to be heritable in cattle [109]. Therefore, if a strong relationship was found between the number of hair follicles per unit area and susceptibility to DD, it might be possible to select for animals that had more or less hair follicles, as appropriate.

5.5. Immune Response

The immune response seems an obvious candidate to examine when looking for sources of variation in susceptibility to a bacterial disease, for example variation in susceptibility to mastitis has been linked to variation in a number of different aspects of the immune response (summarised by [78]). Unfortunately, it seems that dairy cows do not produce an effective immune response to DD: once a cow has had DD, it is significantly more likely to get it again and previous exposure of heifers to an environment where DD is endemic does not appear to reduce their risk of developing the disease after their first calving [71]. Although there has not been much work to date linking susceptibility to DD with variation in the immune response, work is ongoing to develop a better understanding of the immune response to DD, which may point the way to aspects that could influence variation in susceptibility.

There is clear evidence that animals do produce antibodies against treponemes that are associated with DD infection. For example, Walker *et al.* [110] found that levels of antibodies to two spirochaetes associated with PDD were significantly higher in animals with PDD lesions than in animals with no PDD lesions on a California dairy farm. Vink *et al.* [111] developed an ELISA using five Treponema antigens, including representatives from each of the three main serogroups identified in a survey of DD lesions in the UK, and they also found that lesion positive animals had higher antibody titres than lesion negative animals. Recently, Gomez *et al.* [42] developed an ELISA to detect anti-*Treponema* IgG antibodies and found that blood titres of this antibody in heifers increased by an average of 56% after a diagnosis of a first clinical case of DD. The antibody levels decreased to baseline levels within six months for heifers who were treated and did not have a recurrence of the disease, but in heifers with recurring lesions the antibody levels remained high [42].

Although antibodies are produced, the acquired immune response does not seem effective at preventing re-infection with DD [71]. Therefore it may be that if there are any immune differences in susceptibility, they are more likely to be linked to the innate immune response. A poorer innate immune response has been associated with increased susceptibility to infectious diseases in general [76]. Indeed, a number of the SNPs found by Scholey *et al.* [88] to differ between susceptible and non-susceptible animals were found to be in genes related to the inflammatory response, part of the innate immune system.

Another area where individual differences could affect DD susceptibility is in the extent to which the host immune responses are modulated by DD associated treponemes. Recent research has begun to characterise the interactions between treponemes and host tissues within DD lesions. Scholey *et al.* [105] used RNA-Seq technology to compare gene expression in skin biopsies from DD lesions and healthy tissue in Holstein-Friesian dairy cows. It was found that within the lesions there was altered expression of a number of genes relating to immune function and this led the authors to hypothesise that these changes could facilitate the development of DD by reducing the efficacy of the immune response [105]. Refaai *et al.* [112] also examined samples from various DD lesion stages and from healthy skin. They found that the transcription of Interleukin-8 (IL8), a chemokine, was upregulated in the keratinocytes of samples from M2 and M3 lesion stages. IL8 is involved in the inflammatory response and also in keratinocyte proliferation, as seen in the skin condition psoriasis [113], hence it is suggested that its upregulation could help to explain the hyperplasia and thickening of the epidermis observed in DD lesions [112]. As variants of IL8 receptors and promoters have been identified and associated with susceptibility to other infectious diseases, Refaai *et al.* [112] suggest that the role of IL8 in the pathogenesis of DD should be further investigated.

5.6. Behaviour

It is possible that differences in the behaviour of dairy cows could affect their susceptibility to DD. Differences in dairy cow behaviour have already been linked to susceptibility to non-infectious conditions such as claw horn lesions [114,115]. However, as DD is an infectious disease, the ways in which differences in behaviour could affect susceptibility may be different to non-infectious conditions; differences in behaviour which might affect the incidence of cuts and abrasions on the heels, the amount of contact with infected animals, or the amount of contact with reservoirs of infection could all be significant.

Only one study to date has specifically examined dairy cow behaviour and susceptibility to DD. Palmer and O'Connell [116] monitored Holstein-Friesian dairy cows in a cubicle house with solid concrete floors, comparing the behaviour of those that subsequently developed DD with those that did not. They found two sets of behaviours that were performed more by animals that went on to develop DD; standing and walking in the passageway and crosswalks (short passages between the passageway and feed face), and standing half in the cubicle. The authors suggested that more time spent in the passageways could increase the duration of contact between the heels and slurry, possibly increase interactions with infectious animals and may also increase the amount of conflicts with other animals, which could lead to damage to the heel skin. An increased amount of time spent standing half in a cubicle or 'perching' has been associated with the development of lameness and claw horn lesions in previous research [114,115,117]. It has been suggested that standing half in a cubicle could reduce heel height, predisposing animals to infection [114] and it is also likely to increase the contact time between the rear hooves and both slurry and concrete [117]. In addition to behaviour within the housing area of dairy cows, the amount of time that they spend in contact with slurry when standing waiting to be milked could also be of importance. The duration of time spent waiting to be milked can vary greatly between animals within a herd, depending on their position in the milking order.

5.7. Coping Strategies, Personality and Social Status

Individual susceptibility to disease, including DD, might also be affected by how an animal deals with social stress induced by housing and management practices [118]. How animals are affected by a physiological stress response, including effects on the immune system, can vary depending on their personality, coping strategy and social status (reviewed by [118]). It appears that individuals with certain personality types are at higher risk of developing particular pathologies due to the nature of their physiological stress responses [118]. For example, Cohen *et al.* [119] found that people who were more sociable were less susceptible to infection by the common cold virus. The relationship between personality, immune system function and disease susceptibility is complex, however, due to the many processes acting concurrently on an individual [120].

The majority of the literature examining relationships between coping style and physiological stress responses refers to research carried out in laboratory rodents, with a small amount also carried out in pigs [121]. Two differing coping styles (or strategies) are described; a more proactive style associated with a high sympathetic reactivity to stressors, and a more passive (or reactive) style which is typically associated with higher hypothalamic–pituitary–adrenal axis reactivity [121]. Although there is not a great deal of evidence for the relationship between coping styles and susceptibility to disease, the two coping styles have been associated with differences in immune response in pigs [122] and with different immune responses to, followed by differing susceptibility to, an experimentally induced autoimmune disease in rats [123].

The relationship between social status and immunity appears to be linked to exposure to social stressors; for example individuals in a stable group who experience more defeats during social conflicts (and are therefore considered to have low status) have a poorer immune response and are more susceptible to infectious disease [118]. Social status is a less reliable method of categorising animals than personality or coping strategy, however, as social status can vary between contexts [118]. Although research into the relationships between personality types, coping styles, social status and susceptibility to disease has not yet provided clear answers, it is an area that could provide interesting insights into individual differences in susceptibility to different diseases including DD.

6. CONCLUSIONS

Overall, considerable progress has been made in DD research over recent decades, particularly in key areas such as the identification of causative agents and reservoirs of infection. There is, however, still no 'cure' or effective vaccine for DD, so the problems of reduced animal welfare and reduced farm profits remain. A number of factors have been identified which affect DD risk at a farm, animal and individual level. The research into farm level risk factors has identified higher risk housing and management practices, and the research into animal level factors has identified groups of animals which are at higher risk of infection. This information could help farmers to alter their management (where possible) in order to reduce the prevalence of DD on their farms. The number of interacting risk factors at different levels and the infectious nature of DD, however, mean that the factors which are most important could vary between farms and over time.

In terms of factors affecting individual susceptibility, there is some evidence that differences in hoof conformation and behaviour might affect susceptibility to DD. The role that differences in properties of the skin and density of hair follicles could play in susceptibility to DD infection may become clearer as more is learned about the mechanisms of DD transmission between animals and the establishment of DD infections. Investigating the immune response to DD is a complex task due to the variety of bacteria that have been isolated from lesions and the finding that at least some of the DD associated treponemes can modify the host immune response [105,112]. Individual differences in the immune response have been found to affect susceptibility to a number of other infectious diseases in dairy cows [76], and a number of the SNPs identified as differing between DD susceptible and non-susceptible animals by Scholey *et al.* [88] were in regions related to inflammation, therefore it seems likely that immune differences in susceptibility to DD may be detected in the future.

Further work is required to identify the genetic basis of individual variation in susceptibility, to clarify the level of heritability of DD susceptibility and to determine how this is correlated with production and health traits currently used in breeding programmes. Research into the effects of social stressors on animals with different personality types, coping strategies and positions in the social hierarchy could help to inform management strategies which might reduce susceptibility to infectious diseases in general [118]. In addition to this, research to confirm the role of behavioural differences in susceptibility to DD could lead to work examining whether changes in the design of housing might reduce DD prevalence by not only aiming to provide a drier and more hygienic environment for the feet, but also by reducing the proportion of animals standing in passageways and half in cubicles (behaviours found to be associated with DD infection by Palmer and O'Connell [116]). Research into phenotypic and genetic differences between susceptible and non-susceptible individuals, combined with the results of continuing research into the etiology and transmission of the disease, could help in the development of breeding and herd management strategies aimed at reducing the prevalence of DD.

ACKNOWLEDGMENTS

M. Palmer acknowledges the Department for Agriculture for Northern Ireland (DARDNI) for a postgraduate scholarship.

AUTHOR CONTRIBUTIONS

M.P. wrote the review, N.O.C. reviewed and edited.

CONFLICTS OF INTEREST

The authors declare no conflict of interest.

REFERENCES AND NOTES

1. Laven, R.; Proven, M. Use of an antibiotic footbath in the treatment of bovine digital dermatitis. Vet. Rec. 2000, 147, 503–506.

2. Laven, R.; Logue, D. Treatment strategies for digital dermatitis for the UK. Vet. J. 2006, 171, 79–88.

3. Laven, R. Control of digital dermatitis in cattle. In Pract. 2001, 23, 336–341.

4. Sullivan, L.E.; Carter, S.D.; Blowey, R.; Duncan, J.S.; Grove-White, D.; Evans, N.J. Digital dermatitis in beef cattle. Vet. Rec. 2013, 173, 582.

5. Duncan, J.; Angell, J.; Carter, S.; Evans, N.; Sullivan, L.; Grove-White, D. Contagious ovine digital dermatitis: An emerging disease. Vet. J. 2014, 201, 265–268.

6. Sullivan, L.E.; Evans, N.J.; Clegg, S.R.; Carter, S.D.; Horsfield, J.E.; Grove-White, D.; Duncan, J.S. Digital dermatitis treponemes associated with a severe foot disease in dairy goats. Vet. Rec. 2015, 176, 283–287.

7. Clegg, S.R.; Mansfield, K.G.; Newbrook, K.; Sullivan, L.E.; Blowey, R.W.; Carter, S.D.; Evans, N.J. Isolation of Digital Dermatitis Treponemes from Hoof Lesions in Wild North American Elk (Cervus elaphus) in Washington State, USA.J. Clin. Microbiol. 2015, 53, 88–94.

8. Evans, N.; Blowey, R.; Timofte, D.; Isherwood, D.; Brown, J.; Murray, R.; Paton, R.; Carter, S. Association between bovine digital dermatitis treponemes and a range of 'non-healing' bovine hoof disorders. Vet. Rec. 2011, 168, 214–217.

9. Walker, R.; Read, D.; Loretz, K.; Nordhausen, R. Spirochetes isolated from dairy cattle with papillomatous digital dermatitis and interdigital dermatitis. Vet. Microbiol. 1995, 47, 343–355.

10. Choi, B.; Nattermann, H.; Grund, S.; Haider, W.; Gobel, U. Spirochetes from digital dermatitis lesions in cattle are closely related to treponemes associated with human periodontitis. Int. J. Syst. Bacteriol. 1997, 47, 175–181.

11. Cruz, C.E.; Pescador, C.A.; Nakajima, Y.; Driemeier, D. Immunopathological investigations on bovine digital epidermitis. Vet. Rec. 2005, 157, 834–840.

12. Dhawi, A.; Hart, C.; Demirkan, I.; Davies, I.; Carter, S. Bovine digital dermatitis and severe virulent ovine foot rot: a common spirochaetal pathogenesis. Vet. J. 2005, 169, 232–241.

13. Evans, N.J.; Brown, J.M.; Demirkan, I.; Murray, R.D.; Vink, W.D.; Blowey, R.W.; Hart, C.A.; Carter, S.D. Three unique groups of spirochetes isolated from digital dermatitis lesions in UK cattle. Vet. Microbiol. 2008, 130, 141–150.

14. Krull, A.C.; Shearer, J.K.; Gorden, P.J.; Cooper, V.L.; Phillips, G.J.; Plummer, P.J. Deep Sequencing Analysis Reveals Temporal Microbiota Changes Associated with Development of Bovine Digital Dermatitis. Infect. Immun. 2014, 82, 3359–3373.

15. Zinicola, M.; Lima, F.; Lima, S.; Machado, V.; Gomez, M.; Doepfer, D.; Guard, C.; Bicalho, R. Altered Microbiomes in Bovine Digital Dermatitis Lesions, and the Gut as a Pathogen Reservoir. PLoS ONE 2015, 10, e0120504.

16. Read, D.H.; Walker, R.L. Experimental transmission of papillomatous digital dermatitis (footwarts) in dairy cattle. In Proceedings of the 10th International Symposium on Lameness in Ruminants, Casino Lucerne, Switzerland, 7–10 September 1998; p. 270.

17. Gomez, A.; Cook, N.B.; Bernardoni, N.D.; Rieman, J.; Dusick, A.F.; Hartshorn, R.; Socha, M.T.; Read, D.H.; Doepfer, D. An experimental infection model to induce digital dermatitis infection in cattle. J. Dairy Sci. 2012, 95, 1821–1830.

18. Evans, N.J.; Timofte, D.; Isherwood, D.R.; Brown, J.M.; Williams, J.M.; Sherlock, K.; Lehane, M.J.; Murray, R.D.; Birtles, R.J.; Hart, C.A. Host and environmental reservoirs of infection for bovine digital dermatitis treponemes. Vet. Microbiol. 2012, 156, 102–109.

19. Klitgaard, K.; Nielsen, M.W.; Ingerslev, H.; Boye, M.; Jensen, T.K. Discovery of Bovine Digital Dermatitis-Associated Treponema spp. in the Dairy Herd Environment by a Targeted Deep-Sequencing Approach. Appl. Environ. Microbiol.2014, 80, 4427–4432.

20. Rodriguez-Lainz, A.; Hird, D.; Walker, R.; Read, D. Papillomatous digital dermatitis in 458 dairies. J. Am. Vet. Med. Assoc. 1996, 209, 1464–1467. [PubMed]

21. Somers, J.; Frankena, K.; Noordhuizen-Stassen, E.; Metz, J. Risk factors for digital dermatitis in dairy cows kept in cubicle houses in The Netherlands. Prev. Vet. Med. 2005, 71, 11–21.

22. Refaai, W.; Van Aert, M.; Abd El-Aal, A.M.; Behery, A.E.; Opsomer, G. Infectious diseases causing lameness in cattle with a main emphasis on digital dermatitis (Mortellaro disease). Livestock Sci. 2013, 156, 53–63.

23. Laven, R. The environment and digital dermatitis. Cattle Practice 1999, 7, 349–354.

24. Capion, N.; Boye, M.; Ekstrom, C.T.; Jensen, T.K. Infection dynamics of digital dermatitis in first-lactation Holstein cows in an infected herd. J. Dairy Sci. 2012, 95, 6457–6464.

25. Gomez, A.; Cook, N.; Socha, M.; Döpfer, D. First-lactation performance in cows affected by digital dermatitis during the rearing period. J. Dairy Sci. 2015, 98, 4487–4498.

26. Cheli, R.; Mortellaro, C. Digital dermatitis in cattle. In Proceedings of 8th International Conference on Diseases of Cattle, Milan, Italy, 1974; pp. 208–213.

27. Demirkan, I.; Murray, R.; Carter, S. Skin diseases of the bovine digit associated with lameness. Vet. Bull. 2000, 70, 149–171.

28. Holzhauer, M.; Bartels, C.J.M.; Döpfer, D.; van Schaik, G. Clinical course of digital dermatitis lesions in an endemically infected herd without preventive herd strategies. Vet. J. 2008, 177, 222–230.

29. Döpfer, D.; Koopmans, A.; Meijer, F.A.; Szakall, I.; Schukken, Y.H.; Klee, W.; Bosma, R.B.; Cornelissevan, J.L.; Asten, A.J.; ter Huurne, A.A.H.M. Histological and bacteriological evaluation of digital dermatitis in cattle, with special reference to spirochaetes and Campylobacter faecalis. Vet. Rec. 1997, 140, 620–623.

30. Döpfer, D.; Holzhauer, M.; van Boven, M. The dynamics of digital dermatitis in populations of dairy cattle: Model-based estimates of transition rates and implications for control. Vet. J. 2012, 193, 648–653.

31. Berry, S.L.; Read, D.H.; Famula, T.R.; Mongini, A.; Döpfer, D. Long-term observations on the dynamics of bovine digital dermatitis lesions on a California dairy after topical treatment with lincomycin HCl. Vet. J. 2012, 193, 654–658.

32. Read, D.; Walker, R. Papillomatous digital dermatitis (footwarts) in California dairy cattle: clinical and gross pathologic findings. J. Vet. Diagnostic Investig. 1998, 10, 67–76.

33. Demirkan, I.; Walker, R.; Murray, R.; Blowey, R.; Carter, S. Serological evidence of spirochaetal infections associated with digital dermatitis in dairy cattle. Vet. J. 1999, 157, 69–77.

34. Evans, N.J.; Brown, J.M.; Demirkan, I.; Singh, P.; Getty, B.; Timofte, D.; Vink, W.D.; Murray, R.D.; Blowey, R.W.; Birtles, R.J.; et al. Association of Unique, Isolated Treponemes with Bovine Digital Dermatitis Lesions. J. Clin. Microbiol. 2009, 47, 689–696.

35. Evans, N.J.; Brown, J.M.; Scholey, R.; Murray, R.D.; Birtles, R.J.; Hart, C.A.; Carter, S.D. Differential inflammatory responses of bovine foot skin fibroblasts and keratinocytes to digital dermatitis treponemes. Vet. Immunol. Immunopathol. 2014, 161, 12–20.

36. Blowey, R.; Carter, S.; White, A.; Barnes, A. Borrelia burgdorferi infections in UK cattle: A possible association with digital dermatitis. Vet. Rec. 1994, 135, 577–578. [PubMed]

37. Edwards, A.; Dymock, D.; Jenkinson, H. From tooth to hoof: treponemes in tissue-destructive diseases. J. Appl. Microbiol. 2003, 94, 767–780.

38. Antal, G.M.; Lukehart, S.A.; Meheus, A.Z. The endemic treponematoses. Microb. Infect. 2002, 4, 83–94.

39. Collighan, R.J.; Woodward, M.J. Spirochaetes and other bacterial species associated with bovine digital dermatitis.FEMS Microbiol. Lett. 1997, 156, 37–41.

40. Shibahara, T.; Ohya, T.; Ishii, R.; Ogihara, Y.; Maeda, T.; Ishikawa, Y.; Kadota, K. Concurrent spirochaetal infections of the feet and colon of cattle in Japan. Aust. Vet. J. 2002, 80, 497–502.

41. Sullivan, L.E.; Blowey, R.W.; Carter, S.D.; Duncan, J.S.; Grove-White, D.H.; Page, P.; Iveson, T.; Angell, J.W.; Evans, N.J. Presence of digital dermatitis treponemes on cattle and sheep hoof trimming equipment. Vet. Rec. 2014, 175, 201–205.

42. Gomez, A.; Anklam, K.S.; Cook, N.B.; Rieman, J.; Dunbar, K.A.; Cooley, K.E.; Socha, M.T.; Doepfer, D. Immune response against Treponema spp. and ELISA detection of digital dermatitis. J. Dairy Sci. 2014, 97, 4864–4875.

43. Döpfer, D.; Anklam, K.; Mikheil, D.; Ladell, P. Growth curves and morphology of three Treponema subtypes isolated from digital dermatitis in cattle. Vet. J. 2012, 193, 685–693.

44. Keil, D.J.; Liem, A.; Stine, D.L.; Anderson, G.A. Serological and clinical response of cattle to farm specific digital dermatitis bacterins. In Proceedings of the 12th International Symposium on Lameness in Ruminants, Orlando, FL, USA, 9–13 January 2002; p. 385.

45. Berry, S.L.; Ertze, R.A.; Read, D.H.; Hird, D.W. Field evaluation of prophylactic and therapeutic effects of a vaccine against (papillomatous) digital dermatitis of dairy cattle in two California dairies. In Proceedings of the 13th International Symposium on Ruminant Lameness, Maribor, Slovenia, 11–15 February 2003.

46. Berry, S.; Read, D.; Walker, R.; Hird, D.; Shearer, J. Etiology, treatment and prospects for vaccination against (papillomatous) digital dermatitis. In Proceedings of the 12th International Symposium on Lameness in Ruminants, Orlando, FL, USA, 2002; pp. 5–11.

47. Fidler, A.P.; Alley, M.L.; Smith, G.W. Evaluation of a Serpens species bacterin for treatment of digital dermatitis in dairy cattle. Res. Vet. Sci. 2012, 93, 1258–1260.

48. Staton, G.J.; Ainsworth, S.; Blowey, R.W.; Evans, N.J. Developing a vaccine for digital dermatitis of cattle and sheep: A reverse vaccinology approach. In Proceedings of the Cattle Lameness Conference, Worcester, UK, 7 May 2014; pp. 65–66.

49. Manske, T.; Hultgren, J.; Bergsten, C. Prevalence and interrelationships of hoof lesions and lameness in Swedish dairy cows. Prev. Vet. Med. 2002, 54, 247–263.

50. Frankena, K.; Somers, J.G.C.J.; Schouten, W.G.P.; van Stek, J.V.; Metz, J.H.M.; Stassen, E.N.; Graat, E.A.M. The effect of digital lesions and floor type on locomotion score in Dutch dairy cows. Prev. Vet. Med. 2009, 88, 150–157.

51. Hassall, S.A.; Ward, W.R.; Murray, R.D. Effects of lameness on the behavior of cows during the summer. Vet. Rec.1993, 132, 578–580.

52. Margerison, J.K.; Winkler, B.; Stephens, G. The effect locomotion score and lameness and on dry matter intake, feeding and general behaviour. In Proceedings of the 12th Annual Symposium on Lameness in Ruminants, Orlando, FL, USA, 9–13 January 2002.

53. Juarez, S.; Robinson, P.; DePeters, E.; Price, E. Impact of lameness on behavior and productivity of lactating Holstein cows. Appl. Anim. Behav. Sci. 2003, 83, 1–14.

54. Walker, S.; Smith, R.; Jones, D.; Routly, J.; Dobson, H. Chronic stress, hormone profiles and estrus intensity in dairy cattle. Horm. Behav. 2008, 53, 493–501.

55. Almeida, P.; Weber, P.; Burton, J.; Zanella, A. Depressed DHEA and increased sickness response behaviors in lame dairy cows with inflammatory foot lesions. Domest. Anim. Endocrinol. 2008, 34, 89–99.

56. Palmer, M.A.; Law, R.; O'Connell, N.E. Relationships between lameness and feeding behaviour in cubicle-housed Holstein–Friesian dairy cows. Appl. Anim. Behav. Sci. 2012, 140, 121–127.

57. Amory, J.R.; Barker, Z.E.; Wright, J.L.; Mason, S.A.; Blowey, R.W.; Green, L.E. Associations between sole ulcer, white line disease and digital dermatitis and the milk yield of 1824 dairy cows on 30 dairy cow farms in England and Wales from February 2003–November 2004. Prev. Vet. Med. 2008, 83, 381–391.

58. Argaez-Rodriguez, F.; Hird, D.; deAnda, J.; Read, D.; RodriguezLainz, A. Papillomatous digital dermatitis on a commercial dairy farm in Mexicali, Mexico: Incidence and effect on reproduction and milk production. Prev. Vet. Med.1997, 32, 275–286.

59. Hernandez, J.; Shearer, J.; Webb, D. Effect of lameness on milk yield in dairy cows. J. Am. Vet. Med. Assoc. 2002, 220, 640–644.

60. Warnick, L.; Janssen, D.; Guard, C.; Gröhn, Y. The effect of lameness on milk production in dairy cows. J. Dairy Sci.2001, 84, 1988–1997.

61. Relun, A.; Lehebel, A.; Chesnin, A.; Guatteo, R.; Bareille, N. Association between digital dermatitis lesions and test-day milk yield of Holstein cows from 41 French dairy farms. J. Dairy Sci. 2013, 96, 2190–2200.

62. Pavlenko, A.; Bergsten, C.; Ekesbo, I.; Kaart, T.; Aland, A.; Lidfors, L. Influence of digital dermatitis and sole ulcer on dairy cow behaviour and milk production. Animal 2011, 5, 1259–1269.

63. Melendez, P.; Bartolome, J.; Archbald, L.; Donovan, A. The association between lameness, ovarian cysts and fertility in lactating dairy cows. Theriogenology 2003, 59, 927–937.

64. Alawneh, J.; Laven, R.; Stevenson, M. The effect of lameness on the fertility of dairy cattle in a seasonally breeding pasture-based system. J. Dairy Sci. 2011, 94, 5487–5493.

65. Morris, M.; Kaneko, K.; Walker, S.; Jones, D.; Routly, J.; Smith, R.; Dobson, H. Influence of lameness on follicular growth, ovulation, reproductive hormone concentrations and estrus behavior in dairy cows. Theriogenology 2011, 76, 658–668.

66. Rodriguez-Lainz, A.; David, W.; Carpenter, T.; Read, D. Case-control study of papillomatous digital dermatitis in southern California dairy farms. Prev. Vet. Med. 1996, 28, 117–131.

67. Rodriguez-Lainz, A.; Melendez-Retamal, P.; Hird, D.; Read, D.; Walker, R. Farm- and host-level risk factors for papillomatous digital dermatitis in Chilean dairy cattle. Prev. Vet. Med. 1999, 42, 87–97.

68. Wells, S.; Garber, L.; Wagner, B. Papillomatous digital dermatitis and associated risk factors in US dairy herds. Prev. Vet. Med. 1999, 38, 11–24.

69. Holzhauer, M.; Hardenberg, C.; Bartels, C.; Frankena, K. Herd- and cow-level prevalence of digital dermatitis in the Netherlands and associated factors. J. Dairy Sci. 2006, 89, 580–588.

70. Onyiro, O.M.; Andrews, L.J.; Brotherstone, S. Genetic parameters for digital dermatitis and correlations with locomotion, production, fertility traits, and longevity in Holstein-Friesian dairy cows. J. Dairy Sci. 2008, 91, 4037–4046.

71. Laven, R. Determination of the Factors Affecting the Cause, Prevalence and Severity of Digital Dermatitis as a Major Cause of Lameness in Dairy Cows; Milk Development Council Study 95/R1/11; Milk Development Council: Cirencester, UK, 2000; pp. 1–45.

72. Relun, A.; Lehebel, A.; Bruggink, M.; Bareille, N.; Guatteo, R. Estimation of the relative impact of treatment and herd management practices on prevention of digital dermatitis in French dairy herds. Prev. Vet. Med. 2013, 110, 558–562.

73. Barker, Z.E.; Leach, K.A.; Whay, H.R.; Bell, N.J.; Main, D.C.J. Assessment of lameness prevalence and associated risk factors in dairy herds in England and Wales. J. Dairy Sci. 2010, 93, 932–941.

74. Gomez, A.; Bernardoni, N.; Rieman, J.; Dusick, A.; Hartshorn, R.; Read, D.H.; Socha, M.T.; Cook, N.B.; Doepfer, D. A randomized trial to evaluate the effect of a trace mineral premix on the incidence of active digital dermatitis lesions in cattle. J. Dairy Sci. 2014, 97, 6211–6222.

75. Vermunt, J.; Greenough, P. Structural characteristics of the bovine claw: horn growth and wear, horn hardness and claw conformation. Br. Vet. J. 1995, 151, 157–180.

76. Detilleux, J.C.; Koehler, K.J.; Freeman, A.; Kehrli, M.E.; Kelley, D.H. Immunological Parameters of Periparturient Holstein Cattle: Genetic Variation. J. Dairy Sci. 1994, 77, 2640–2650.

77. Enevoldsen, C.; Grohn, Y.T.; Thysen, I. Skin injuries on the body and thigh of dairy cows: associations with season, claw health, disease treatment, and other cow characteristics. Acta Vet. Scand. 1994, 35, 337–347. [PubMed]

78. Detilleux, J. Genetic factors affecting susceptibility of dairy cows to udder pathogens. Vet. Immunol. Immunopathol.2002, 88, 103–110.

79. Benjamin, A.; Green, B.; Hayden, L.; Barlow, J.; Kerr, D. Cow-to-cow variation in fibroblast response to a toll-like receptor 2/6 agonist and its relation to mastitis caused by intramammary challenge with Staphylococcus aureus. J. Dairy Sci. 2015, 98, 1836–1850.

80. Pinedo, P.; Galvão, K.; Seabury, C. Innate immune gene variation and differential susceptibility to uterine diseases in Holstein cows. Theriogenology 2013, 80, 384–390.

81. Fisher, C.A.; Bhattarai, E.K.; Osterstock, J.B.; Dowd, S.E.; Seabury, P.M.; Vikram, M.; Whitlock, R.H.; Schukken, Y.H.; Schnabel, R.D.; Taylor, J.F. Evolution of the bovine TLR gene family and member associations with Mycobacterium avium subspecies paratuberculosis infection. PLoS ONE 2011, 6, e27744.

82. Pinedo, P.J.; Buergelt, C.D.; Donovan, G.A.; Melendez, P.; Morel, L.; Wu, R.; Langaee, T.Y.; Rae, D.O. Candidate gene polymorphisms

(BoIFNG, TLR4, SLC11A1) as risk factors for paratuberculosis infection in cattle. Prev. Vet. Med. 2009,91, 189–196.

83. Rupp, R.; Boichard, D. Genetic parameters for clinical mastitis, somatic cell score, production, udder type traits, and milking ease in first lactation Holsteins. J. Dairy Sci. 1999, 82, 2198–2204.

84. Kelm, S.; Dettilleux, J.; Freeman, A.; Kehrli, M.; Dietz, A.; Fox, L.; Butler, J.; Kasckovics, I.; Kelley, D. Genetic association between parameters of innate immunity and measures of mastitis in periparturient Holstein cattle. J. Dairy Sci. 1997, 80, 1767–1775.

85. van der Waaij, E.; Holzhauer, M.; Ellen, E.; Kamphuis, C.; de Jong, G. Genetic parameters for claw disorders in dutch dairy cattle and correlations with conformation traits. J. Dairy Sci. 2005, 88, 3672–3678.

86. Oberbauer, A.M.; Berry, S.L.; Belanger, J.M.; McGoldrick, R.M.; Pinos-Rodriquez, J.M.; Famula, T.R. Determining the heritable component of dairy cattle foot lesions. J. Dairy Sci. 2013, 96, 605–613.

87. Oikonomou, G.; Cook, N.; Bicalho, R. Sire predicted transmitting ability for conformation and yield traits and previous lactation incidence of foot lesions as risk factors for the incidence of foot lesions in Holstein cows. J. Dairy Sci. 2013, 96, 3713–3722.

88. Scholey, R.A.; Blowey, R.W.; Murray, R.D.; Smith, R.F.; Cameron, J.; Massey, J.P.; Ollier, W.E.; Carter, S.D. Investigating host genetic factors in bovine digital dermatitis. Vet. Rec. 2012, 171, 624.

89. Carter, S. Digital dermatitis—How is it spread and can we stop it? In Proceedings of the Cattle Lameness Conference, Worcester, UK, 7 May 2014; pp. 43–48.

90. Boettcher, P.; Dekkers, J.; Warnick, L.; Wells, S. Genetic analysis of clinical lameness in dairy cattle. J. Dairy Sci. 1998,81, 1148–1156.

91. Walker, R.L.; Berry, S.L.; Rodriquez-Lainz, A.; Read, D. Prospective study on foot conformation characteristics predisposing to the development of papillomatous digital dermatitis. In Proceedings of the 12th International Symposium on Lameness in Ruminants, Orlando, FL, USA, 9–13 January 2002; p. 370.

92. Laven, R. The relationship between hoof conformation and digital dermatitis in dairy cattle. Cattle Practice 2007, 15, 93–95.

93. Olechnowicz, J.; Jaskowski, J.M. Hoof Measurements Related to Locomotion Scores and Claw Disorders in Dairy Primiparous Cows. Bull. Vet. Inst. Pulawy 2010, 54, 87–92.

94. Baird, L.G.; O'Connell, N.E.; McCoy, M.A.; Keady, T.W.J.; Kilpatrick, D.J. Effects of breed and production system on lameness parameters in dairy cattle. J. Dairy Sci. 2009, 92, 2174–2182.

95. Daniel, V. Digital dermatitis and inter-digital dermatitis infection risks increase as inter-digital cleft space in Holsteins decrease. In Proceedings of the 16th International Symposium and 8th International Conference on Lameness in Ruminants, Rotorua, New Zealand, 28 February–3 March 2011; p. 14.

96. Gomez, A.; Cook, N.B.; Rieman, J.; Dunbar, K.A.; Cooley, K.E.; Socha, M.T.; Doepfer, D. The effect of digital dermatitis on hoof conformation. J. Dairy Sci. 2015, 98, 927–936.

97. Palmer, M.A.; Donnelly, R.F.; Garland, M.J.; Majithiya, R.; O'Connell, N.E. The effect of slurry on skin permeability to methylene blue dye in dairy cows with and without a history of digital dermatitis. Animal 2013, 7, 1731–1737.

98. Holbrook, K.A.; Odland, G.F. Regional differences in the thickness (cell layers) of the human stratum corneum: an ultrastructural analysis. J. Invest. Dermatol. 1974, 62, 415–422.

99. Elias, P.M.; Cooper, E.R.; Korc, A.; Brown, B.E. Percutaneous transport in relation to stratum corneum structure and lipid composition. J. Invest. Dermatol. 1981, 76, 297–301.

100. Palmer, C.N.; Irvine, A.D.; Terron-Kwiatkowski, A.; Zhao, Y.; Liao, H.; Lee, S.P.; Goudie, D.R.; Sandilands, A.; Campbell, L.E.; Smith, F.J. Common loss-of-function variants of the epidermal barrier protein filaggrin are a major predisposing factor for atopic dermatitis. Nat. Genet. 2006, 38, 441–446.

101. Jung, T.; Stingl, G. Atopic dermatitis: therapeutic concepts evolving from new pathophysiologic insights. J. Allergy Clin. Immunol. 2008, 122, 1074–1081.

102. Barry, B. Drug delivery routes in skin: a novel approach. Adv. Drug Deliv. Rev. 2002, 54, S31–S40.

103. Knorr, F.; Lademann, J.; Patzelt, A.; Sterry, W.; Blume-Peytavi, U.; Vogt, A. Follicular transport route—Research progress and future perspectives. European J. Pharmaceut. Biopharmaceut. 2009, 71, 173–180.

104. Wosicka, H.; Cal, K. Targeting to the hair follicles: current status and potential. J. Dermatol. Sci. 2010, 57, 83–89.

105. Scholey, R.; Evans, N.; Blowey, R.; Massey, J.; Murray, R.; Smith, R.; Ollier, W.; Carter, S. Identifying host pathogenic pathways in bovine digital dermatitis by RNA-Seq analysis. Vet. J. 2013, 197, 699–706.

106. Lademann, O.; Kramer, A.; Richter, H.; Patzelt, A.; Meinke, M.; Roewert-Huber, J.; Czaika, V.; Weltmann, K.; Hartmann, B.; Koch, S. Antisepsis of the follicular reservoir by treatment with tissue-tolerable plasma (TTP). Laser Phys. Lett. 2011, 8, 313.

107. Ossadnik, M.; Czaika, V.; Teichmann, A.; Sterry, W.; Tietz, H.; Lademann, J.; Koch, S. Differential stripping: introduction of a method to show the penetration of topically applied antifungal substances into the hair follicles.Mycoses 2007, 50, 457–462.

108. Bertipaglia, E.; Silva, R.; Maia, A. Fertility and hair coat characteristics of Holstein cows in a tropical environment.Anim. Reprod. 2005, 2, 187–194.

109. Bertipaglia, E.C.A.; da Silva, R.G.; Cardoso, V.; Fries, L.A. Hair coat characteristics and sweating rate of Braford cows in Brazil. Livestock Sci. 2007, 112, 99–108.

110. Walker, R.; Read, D.; Loretz, K.; Hird, D.; Berry, S. Humoral response of dairy cattle to spirochetes isolated from papillomatous digital dermatitis lesions. Am. J. Vet. Res. 1997, 58, 744–748. [PubMed]

111. Vink, W.D.; Jones, G.; Johnson, W.O.; Brown, J.; Demirkan, I.; Carter, S.D.; French, N.P. Diagnostic assessment without cut-offs: Application of serology for the modelling of bovine digital dermatitis infection. Prev. Vet. Med. 2009,92, 235–248.

112. Refaai, W.; Ducatelle, R.; Geldhof, P.; Mihi, B.; El-shair, M.; Opsomer, G. Digital dermatitis in cattle is associated with an excessive innate immune response triggered by the keratinocytes. BMC Vet. Res. 2013, 9, 193.

113. Tuschil, A.; Lam, C.; Haslberger, A.; Lindley, I. Interleukin-8 stimulates calcium transients and promotes epidermal cell proliferation. J. Invest. Dermatol. 1992, 99, 294–298.

114. Galindo, F.; Broom, D.; Jackson, P. A note on possible link between behaviour and the occurrence of lameness in dairy cows. Appl. Anim. Behav. Sci. 2000, 67, 335–341.

115. Proudfoot, K.; Weary, D.; Von Keyserlingk, M. Behavior during transition differs for cows diagnosed with claw horn lesions in mid lactation. J. Dairy Sci. 2010, 93, 3970–3978.

116. Palmer, M.A.; O'Connell, N.E. The relationship between dairy cow behaviour in a cubicle house and susceptibility to digital dermatitis. In Proceedings of the Cattle Lameness Conference, Worcester, UK, 2 May 2012; pp. 41–42.

117. Dippel, S.; Tucker, C.B.; Winckler, C.; Weary, D.M. Effects of behaviour on the development of claw lesions in early lactation dairy cows. Appl. Anim. Behav. Sci. 2011, 134, 16–22.

118. Proudfoot, K.L.; Weary, D.M.; von Keyserlingk, M.A. Linking the social environment to illness in farm animals. Appl. Anim. Behav. Sci. 2012, 138, 203–215.

119. Cohen, S.; Doyle, W.J.; Turner, R.; Alper, C.M.; Skoner, D.P. Sociability and susceptibility to the common cold. Psychol. Sci. 2003, 14, 389–395.

120. Friedman, H.S. The multiple linkages of personality and disease. Brain Behav. Immun. 2008, 22, 668–675.

121. Koolhaas, J. Coping style and immunity in animals: making sense of individual variation. Brain Behav. Immun. 2008,22, 662–667.

122. Hessing, M.-J.-C.; Coenen, G.-J.; Vaiman, M.; Renard, C. Individual differences in cell-mediated and humoral immunity in pigs. Vet. Immunol. Immunopathol. 1995, 45, 97–113.

123. Kavelaars, A.; Heijnen, C.J.; Tennekes, R.; Bruggink, J.E.; Koolhaas, J.M. Individual behavioral characteristics of wild-type rats predict susceptibility to experimental autoimmune encephalomyelitis. Brain Behav. Immun. 1999, 13, 279–286.

CHAPTER 8

Farm Management in Organic and Conventional Dairy Production Systems Based on Pasture in Southern Brazil and Its Consequences on Production and Milk Quality

Shirley Kuhnen 1,2, Rudinei Butka Stibuski 2, Luciana Aparecida Honorato 2 and Luiz Carlos Pinheiro Machado Filho1,2,*

1Department of Zootechny and Rural Development, Federal University of Santa Catarina, Florianópolis 88034-001, Brazil

2Graduate Program in Agroecosystems, Federal University of Santa Catarina, Florianópolis 88034-001, Brazil

SIMPLE SUMMARY

This study provides the characteristics of the conventional high input (C-HI), conventional low input (C-LI), and organic low input (O-LI) pasture-based production systems used in Southern Brazil, and its consequences on production and milk quality. C-HI farms had larger farms and herds, annual pasture with higher inputs and milk yield, whereas O-LI had smaller farms and herds, perennial pastures with lowest input and milk yields; C-LI was in between. O-LI farms may contribute to eco-system services, but low milk yield is a major concern. Hygienic and microbiological milk quality was poor for all farms and needs to be improved.

ABSTRACT

Pasture-based dairy production is used widely on family dairy farms in Southern Brazil. This study investigates conventional high input (C-HI), conventional low input (C-LI), and organic low input (O-LI) pasture-based systems and their effects on quantity and quality of the milk produced. We conducted technical site visits and interviews monthly over one year on 24 family farms (n = 8 per type). C-HI farms had the greatest total area (28.9 ha), greatest percentage of area with annual pasture (38.7%), largest number of lactating animals (26.2) and greatest milk yield per cow (22.8 kg·day^{-1}). O-LI farms had the largest perennial pasture area (52.3%), with the greatest botanical richness during all seasons. Area of perennial pasture was positively correlated with number of species consumed by the animals ($R^2 = 0.74$). Milk from O-LI farms had higher levels of fat and total solids only during the winter. Hygienic and microbiological quality of the milk was poor for all farms and need to be improved. C-HI farms had high milk yield related to high input, C-LI had intermediate characteristics and O-LI utilized a year round perennial pasture as a strategy to diminish the use of supplements in animal diets, which is an important aspect in ensuring production sustainability.

KEYWORDS

organic milk; pasture-based dairy production; sustainable agriculture; food security

1. INTRODUCTION

Milk is the top agriculture commodity in monetary value in the world, and Brazil is the world's fourth largest fresh cow milk producer [1]. Within Brazil, Western Santa Catarina State, Southern Brazil, is the third largest and fastest growing milk-producing region [2]. On its way to becoming a major player in milk exports, the region is characterized by diversity in dairy production systems and is dominated by family farming (80% of farms) [3]. The region has a mild, subtropical humid climate, without a dry season and with temperate summers, allowing year round pasture-based management. These pasture systems can consist of either extensive or rotational grazing, and animals may be supplied with different amounts of supplements (concentrate and silage). In semi-intensive systems, cows are fed with large amounts of supplements. On the other hand, free-stall confinement is rare in the region [4].

World organic food production is growing at a rate of 10%–15% annually [5]. The growth of the organic market is likely associated with consumers' concerns related to human health risks (e.g., degenerative diseases) caused by the use of pesticides, antibiotics, hormones and heavy metals in agriculture [6]. Furthermore, environmental impact and animal and human welfare are important considerations in organic farming practices and, therefore, organic production has been adopted by many family dairy farms globally. In Western Santa Catarina State, dairy farmers have been converting to organic production due to the assumed lower production costs while maintaining high levels of performance [4,7]. It is important to note that, regardless of the management system used, maximizing the use of forage to feed ruminants is essential in improving eco-efficiency [8]. Concentrate use increases with dairy farm intensification [8,9]. However, recommendations to improve the environmental impact of dairy farms include reducing the use of concentrates per kg of milk so as to improve the nutrient flow of the farm [9].

Comparative studies between conventional and organic pasture-based production systems are rare in the literature. The majority of studies conducted in the USA and Europe compare organic with free-stall farming, which are different from the dairy systems used in Brazil (e.g., see: [10,11]). Nevertheless, some studies have found differences between systems in relation to feed management, the use of forage species [12,13] and eco-efficiency [8]. Animals that receive a diet based on pasture, compared to a diet based on preserved forage and concentrate supplements, produce milk with greater concentrations of chemical components that are beneficial for human health, including conjugated linoleic acid, carotenoids, lipid-soluble vitamins A and E and antioxidants [10,11]. For consumers, knowledge about differences in milk quality is important for decision making related to food choice; for farmers, knowing what causes differences in food quality and production performance is important information for decision making related to production and management.

Recently, dairy industries have increased their presence in the study region, increasing milk processing capacity [4] and, as a consequence, putting pressure on dairy farmers to increase their production, which entails use of high-yield dairy breeds and a considerable input of feed grain that are known to rapidly increase productivity [14]. Despite increasing production, intensification may also increase costs and have significant environmental impacts [15]. A study providing a detailed assessment of the characteristics of pasture-based production systems used in a typical and growing dairy region may provide knowledge that can be used to inform future applied studies or decision making related to milk quality. Our field study seeks to describe pasture-based dairy production systems used in Southern Brazil and to assess the possible effects of management on diet composition, milk production, composition and quality. To our knowledge, this is the first study comparing production traits and milk quality between conventional and organic pasture-based production systems.

2. MATERIALS AND METHODS

This study was conducted in Santa Catarina State, Southern Brazil, between May 2011 and April 2012. Twenty-four family farms were included in the analysis, and each farm was visited monthly. Based on earlier studies [4,7,16] and information gathered from technical assistants in the region, only pasture-based farms were included. We defined three categories of dairy production systems based on the typical management from that region: Conventional High input Pasture-based (C-HI), Conventional Low input Pasture-based (C-HI), and Organic Low input Pasture-based (O-LI). Eight farms per management type were included in the study. C-HI farms have a high intake of concentrates and preserved forage year round (>10 kg DM/cow/day); C-LI farms have a low intake of concentrates and preserved forage (<10 kg DM/cow/day); and O-LI farms are low input pasture-based dairy farms that are certified to be organic by the Ecovida Network in accordance with Brazilian Legislation [17].

All farms included in the study were located in Santa Catarina, with similar soil and climate characteristics. The climate is humid subtropical (Cfb in the Köppen classification) with an annual rainfall between 1900 and 2200 mm and a mean air temperature of 18 to 20 °C [18]. The farmers have similar ethnic background and use similar agricultural techniques. We used semi-structured interviews to collect data on the following variables on the first visit: total farm area, land use, area for cultivating perennial and annual pasture. During each monthly visit we collected data on breed of animals, number of lactating animals, monthly herd milk yield (obtained from sales records) and diet composition. In assessing nutritional management, we recorded the production of silage (tons/year) and estimated the daily consumption of silage (kg/cow), and daily consumption of concentrate (obtained from sale records) (kg/cow). For the calculation of silage and concentrate dry matter intake, the dry matter concentration in silage and concentrate samples were analyzed according to Van Soest [19]. The silage and concentrate dry matter intake was estimated based on the information given by the producers and dry matter determined from collected samples. We determined the forage species consumed by the animals using the methodology described by Euclides *et al.* [20], which consists of simulated grazing after observing the feeding behavior of animals for 30 min. The sampling of plant species consumed by the animals during observation was performed at each farm and during each season throughout the year.

Due to differences in the breeds used at each farm, and as recommended by other authors [4,7], milk samples for chemical analysis were obtained from the milking of three pre-selected animals based on lactation period (between the 4th and 6th month), phenotype and similar genetic composition (3/4 Holstein and 1/4 Jersey). The total milking of the three selected cows was mixed and a sample was collected for chemical analysis. Further, milk samples from the bulk tank of each farm were collected for the analysis of somatic cell count and bacteria count (SCC and BMBC). The procedure described above was performed twice in each season, on alternate days, and all samples were sent refrigerated to a Laboratory certified by the Ministry of Agriculture. We used infrared spectroscopy to analyze total solids, protein, fat, non-fat solids and lactose (Delta Combiscope e Bentley Combisystem B-2300) [21]. To analyze SCC [22] and BMBC we used flow cytometry (Delta Combiscope and Bentley Combisystem B-2300; Bentley Bactocount IBC) [23].

For statistical analysis, we used a log transformation of the variables BMBC and SCC to meet the conditions of normality and the mixed model (PROC MIXED, SAS version 9.0) for data analysis with repeated measures, with fixed effects for management type (C-HI, C-LI, and O-LI), season (autumn, winter, spring, summer) and their interactions. The subject of repeated measurement was the farm. Of the several error structures investigated, the heterogeneous autoregressive structure (ARH) was selected because this co-variance structure minimized the BIC (Bayes Schwarz information criterion), except for milk production data, where we used unstructured covariance (UN). For presentation purposes, data were transformed back. The effects of the management type for each season were compared using Fisher's least significant difference. In all comparisons, significance was set at $p \leq 0.05$. The studied variables (total farm area, percentage of area used for perennial and annual pasture, number of lactating animals per season, daily milk yield in each season, estimated daily intake of concentrate and silage per season, number of forage species consumed per season) were also submitted to principal component analysis (PCA), using the statistical package Unscrambler 9.0. Preprocessing of raw data included normalizing, and the analysis was conducted using correlation matrices with mean adjusted data.

3. RESULTS AND DISCUSSION

3.1. Farm Characteristics

The management types differed in their productive traits (Table 1). The farms using C-HI management had the largest total area compared to the other management systems and, in general, the area with annual pasture was greater on farms with conventional management (C-HI and C-LI) compared to O-LI. The O-LI farms had the greatest area of perennial pasture. The area designated for the cultivation of perennial pasture in conventional farms (C-HI and C-LI) was only 18.8 and 23.7%, respectively, which differed significantly from O-LI farms (52.3%). In farms with C-HI and C-LI, a large proportion of the farm's area was used for integrated crop-livestock, with a predominance of annual summer crops, mainly soybeans and corn, for grain or silage production (38.7% and 38.1% of the total area for C-HI and C-LI, respectively). In winter, these same areas were used for annual pasture cultivation. Additionally, besides having to follow environmental regulations restricting land use (20% of the farm must be protected as legal reserve area), parts of the productive area of all farms were used for the production of subsistence crops (Table 1), which is characteristic of family farming. C-HI farms had a higher number of lactating cows, with more Holsteins in their herds compared to O-LI farms, with being C-LI in an intermediary position (Table 1). Likewise, the total milk yields were 625 $kg \cdot day^{-1}$, 384 $kg \cdot day^{-1}$ and 130 $kg \cdot day^{-1}$ for C-HI, C-LI and O-LI farms, respectively. The primary reason for this difference is the number of lactating cows in each farm type (26.3, 19.1 and 13.1, respectively, in C-HI, C-LI and O-LI farms), but also the genetics and the feeding of the cows (Table 1and Table 2).

3.2. Diet Composition, Production and Milk Quality

The use of concentrate and silage maize in conventional management occurs throughout the year (Table 2). The constant use of supplements indicates that their use was not exclusively associated with seasonal fluctuation in pasture production; they were also used as a strategy to increase milk yield (Table 3). This contrasts with the O-LI management, where the use of concentrate and/or silage was very minor, occurring only during winter and the beginning of spring. On C-HI farms the cows were left to graze during certain periods of the day, the pasture was managed rotationally and with the use of chemical fertilizers. In C-HI management, milk production relied not only on fertilized pastures, but also on the amount of concentrate and silage fed to the cows, which was greater than that used in other farm types. Thus, the average production was 22.8, 18.1 and 9.6, per cow per day in C-HI, C-LI and O-LI, respectively. The results for C-LI farms showed an intermediary situation between O-LI and C-HI. Although not relying completely on pasture, as O-LI did, they could not afford the same level of inputs as C-HI farmers, due to limited financial resources.

Table 1. Characteristics of pasture-based dairy production systems in Western Santa Catarina State, Southern Brazil, grouped by conventional high input (C-HI), conventional low input (C-LI), and organic low input (O-LI) (n = 8 per group).

Items	Management Types			
	C-HI	C-LI	O-LI	SEM
Total farm area (ha)	28.9 a	21.4 a,b	18.6 b	1.75
Area of perennial pasture (%)	18.8 b	23.7 b	52.3 a	5.98
Area of annual pasture (%)	38.7 a	38.1 a	12.9 b	4.97
Other (building, crops, subsistence and legal reserve area) (%)	42.5	38.2	34.4	8.89
Breed of lactating cows (%)				
Holstein	50.6	33.4	20.8	3.42
Jersey	13.0	20.8	20.1	1.40
Holstein Jersey crossbred	36.4	45.8	59.1	2.99
Number of lactating animals	26.3 a	19.1 b	13.1 c	1.79
Number of lactating cows/ha	1.9 a	1.4 a,b	1.2 b	0.15

SEM = standard error of the mean. Different lower case letters on the same line indicate statistical difference between management types at a level of 5%.

The nutritional management adopted by organic farmers follows Brazilian Legislation [24] in relation to the maximum (15% DM) permitted percentage of conventional supplements that must be gene manipulated organism (GMO)-free. In Brazil, GMO corn and soybean varieties dominate production, making up 90% and 92%, respectively, of all corn and soybean crops [25], and as such they are very likely grown on neighboring farms. GMO-free corn is rarely found in the region. Although some farmers cultivate their own corn, it is unlikely completely "GMO-free", as required by law, due to the proximity to farms with GMO crops. Therefore, milk production in O-LI farms has to rely on the efficient management of pasture. To achieve a high pasture production, all O-LI farms adopted Voisin's rotational grazing (VRG), which is an intensive and time-controlled rotational grazing system, with the aim of providing the nutritional requirements of cows while also allowing pasture regrowth. Pastures are divided into plots [26], and animals are provided with water and shade [27]. In the region, VRG farms typically have 60–80 paddocks. Milking cows stay 12 h in each paddock at an instant stocking rate of 200 cows/ha, followed by dry cows that stay in the paddock for another 12 h; the resting time of each paddock varies from 25 to 50 days, depending on the season. Subdividing large pastures may reduce distances to water and provide more uniform use of forage [28]. With a higher stocking density of animals and a shorter rotation period for each paddock, this system improves animal productivity [28,29].

Table 2. Estimated supplement intake for cows from farms grouped by conventional high input (C-HI), conventional low input (C-LI), and organic low input (O-LI) (n = 8 per group).

Diet Composition * (kg DM/cow/day)	Season	Management Types			
		C-HI	C-LI	O-LI	SDM
Concentrate					
	Autumn	3.83	1.73	0	2.00
	Winter	3.65	1.46	0.20	1.64
	Spring	3.32	1.55	0.10	1.48
	Summer	3.20	1.87	0	1.44
Silage					
	Autumn	8.31	5.38	0	3.91
	Winter	7.66	5.37	1.30	3.43
	Spring	7.04	3.50	0.20	3.32
	Summer	6.33	6.33	0	3.57

* Intake of concentrate and silage was estimated by farmers and dry mass (DM) content was analyzed in concentrate and silage sampled on the farms; Statistical analyses were not performed as data available are estimates.

Despite some similarities between conventional management systems, C-LI farms appear to be intermediary between C-HI and O-LI. This is supported by similarities in the number of lactating animals, breed of animals, herd and cow milk yield and use of supplements in the diet (Table 1 and Table 2). We found that C-LI farms made limited investments in genetically improved breeds, pasture fertilization and the purchasing of large quantities of feed supplement. In relation to the identification of forage species consumed by the animals in each management type, we found that on conventional farms (C-HI and C-LI) common oats (*Avena sativa*) and annual ryegrass (*Lolium multiflorum*) were planted in autumn and winter either as a single crop or intercropped (Table 4). During the other seasons, depending on the intervals between each crop, farmers planted annual pasture in the summer, particularly Sudan grass (*Sorghum sudanensis*) and pearl millet (*Pennisetum americanum*), as a monoculture or intercropped (Table 4).

Table 3. Milk yield, percentage of fat, protein, lactose and total solids in milk from farms grouped by conventional high input (C-HI), conventional low input (C-LI), and organic low input (O-LI) (n = 8 per group) in Western Santa Catarina State, Southern Brazil.

Parameters	Season	Management Types			
		CH-I	CL-I	O-LI	SEM
Milk yield (kg/lactating cow/day) corrected fat 4%	Autumn	23.1 ABa	16.8 b	9.2 c	1.45
	Winter	22.1 Ba	18.0 ab	10.8 c	1.45
	Spring	25.8 Aa	19.2 ab	9.8 c	1.45
	Summer	22.8 ABa	19.3 ab	10.2 c	1.45
Fat (%)	Autumn	4.4 Aa	4.1 Aa	4.4 Ba	0.20
	Winter	4.0 ABa	3.9 Aa	4.8 Ab	0.21
	Spring	4.5 Aa	4.2 Aa	4.0 Ba	0.22
	Summer	3.9 Ba	4.1 Aa	4.1 Ba	0.15
Protein (%)	Autumn	3.4	3.2	3.3	0.10
	Winter	3.7	3.3	3.3	0.25
	Spring	3.2	3.3	3.2	0.07
	Summer	3.1	3.3	3.1	0.06
Lactose (%)	Autumn	4.3	4.3	4.3	0.09
	Winter	4.5	4.4	4.4	0.07
	Spring	4.5	4.4	4.6	0.05
	Summer	4.4	4.4	4.6	0.07
Total Solids (%)	Autumn	13.1 Aa	12.6 Aa	12.9 Aa	0.26
	Winter	11.9 Ba	11.8 Ba	12.6 Ab	0.21
	Spring	13.2 Aa	12.9 Aa	12.7 Aa	0.21
	Summer	12.3 Ba	12.6 Aa	12.5 Aa	0.18

SEM = standard error means between management types. Different lower case letters in the same row indicate statistical differences between management types in the same season; different upper case letters in the same column indicate statistical differences of the same management types between seasons ($p < 0.05$).

Higher levels of fat in organic milk in comparison to conventional milk have also been attributed to the predominant use of Jersey cows, animals in the early stages of lactation and a greater consumption of fresh forage on organic farms [10]. In our study, however, we controlled for the effect of breed and lactation stage by excluding animals that did not fit within the determined breed type (Holstein x Jersey) and animals that were between the 4th and 6th month of lactation.

Pasture management on VRG can explain the greater species richness encountered on the organic farms, and consequently, the greater consumption by the animals (Table 4). Previous studies have demonstrated an increase in the diversity of species through pasture management in comparison to unmanaged fields [30], as well as the positive effect of intensive management on pasture quality. In this case, the pasture achieves higher levels of protein and lower crude fiber than extensively managed pasture [30]. Adler *et al.* [13] also found differences in botanical composition, with a lower proportion of grasses and greater proportion of legumes on organic farms compared to conventional farms.

C-HI and C-LI farmers managed their perennial pastures on a rotational system with fixed resting days, where the cows stayed on pasture for part of the day and in the barn receiving silage and concentrate for the other part of the day. Weeds are controlled with chemicals, and soluble fertilizers are used on the pasture. None of these methods are used on the O-LI farms. The management methods used may explain the absence of legumes found in the cows' diet on C-HI and C-LI farms.

Table 4. Forage species identified in simulated grazing samples on two or more farms grouped by conventional high input (C-HI), conventional low input (C-LI), and organic low input (O-LI) (n = 8 per group) in Southern Brazil.

Season	Management Types	Plant Species		
		Cultivated Grass	Legume	Native Forage
Autumn	C-HI	*Avena sativa*; *Eleusine indica*; *Sorghum sudanensis*		*Urochloa plantaginea*
	C-LI	*Avena sativa*; *Avena strigosa*; *Lolium multiflorum*; *Sorghum sudanensis*		*Cynodon dactylon*
	O-LI	*Axonopus affinis*; *Axonopus campressus*; *Lolium multiflorum*; *Saccharum officinarum*; *Pennisetum purpureum*	*Trifolium repens*	*Ageratum conyzoides*; *Amaranthus deflexus*; *Amaranthus spinosus*; *Kylling abrevifolia*; *Paspalum mandiocanum*; *Paspalum umbrosum*; *Paspalum urvillei*; *Plantago australis*; *Solidago chilensis*
Winter	C-HI	*Avena strigosa*; *Lolium multiflorum*		
	C-LI	*Avena strigosa*; *Lolium multiflorum*		
	O-LI	*Avena strigosa*; *Lolium multiflorum*	*Trifolium repens*	*Bowlesia incana*; *Bromus catharticus*; *Cerastium glomeratum*; *Hypochaeris megapotamica*; *Hypochaeris* sp.; *Paspalum umbrosum*; *Plantago australis*; *Veronica arvensis*
Spring	C-HI	*Avena sativa*; *Lolium multiflorum*		
	C-LI	*Avena sativa*; *Lolium multiflorum*		
	O-LI	*Avena strigosa*; *Lolium multiflorum*	*Trifolium repens*; *Vicia sativa*	*Cynodon defensis*; *Leonurus sibiricus*; *Paspalum urvillei*; *Penisetum setaceum*; *Plantago australis*

Table 4. *Cont.*

Season	Management Types	Plant Species		
		Cultivated Grass	Legume	Native Forage
Summer	C-HI	*Pennisetum americanum*; *Sorghum sudanensis*		
	C-LI	*Pennisetum americanum*; *Sorghum sudanensis*		
	O-LI	*Pennisetum purpureum*; *Sorghum sudanensis*	*Desmodium affine*; *Glicine max*	*Digitaria ciliaris*; *Euphorbia heterophylla*; *Paspalum juergensii*; *Paspalum umbrosum*; *Urochloa plantaginea*

Data obtained based on the methodology described by Euclides *et al.* [20].

We found an inverse correlation between the percentage of the farm area used for annual pasture and lower levels of pasture richness ($r^2 = -0.70$; $p < 0.05$), as well as a positive correlation between the percentage of land used for perennial pasture and species-rich pasture ($r^2 = 0.74$, $p < 0.05$). As can be seen in Table 1, O-LI farms have significantly more perennial pasture than C-HI and C-LI, with greater species richness, which is an indicator of better sustainability and higher productivity [31]. Perennial pasture is related to improved sustainability and productivity because their long-lived deep root systems take up soil water more efficiently, reduce soil acidification and reduce the risk of soil erosion by maintaining plant cover during summer, improve soil structure and water infiltration, and provide feed for livestock when annual pastures have senesced [32]. Pastures made up of only one or two species often have less production stability than pastures with a greater complexity of species [33]. In this sense, increasing the pasture diversity may represent an important strategy for the maintenance of pasture-based animals. Mixed pastures can be more profitable and more productive than monocultures, while also providing less risk of weather-related loss than pastures made up of only one grass species [34].

The milk yield per cow per day was greater in the C-HI farms compared to organic farms in every season throughout the year (Table 3). The lower milk yield per cow seen in O-LI farms is likely associated with the use of mixed breed animals and a diet with low input of supplements, along with smaller farm size and lower number of lactating cows. These characteristics explain the lower milk yield at the level of the herd, which seems to be something that could be improved on organic farms. Previous studies have shown that a large proportion of organic [7,35] and low-input conventional farms (that make limited use of external supplements) tend to breed crosses that provide more hardy animals that are better adapted to local climatic conditions such as temperature and more suitable to hilly terrains typical of the region [4,10,16]. In general, despite lower milk production, farmers have adopted crossbreeds to give more emphasis on traits such as fiber use efficiency, fertility, health and survival in the herd [36].

Higher levels of fat and total solids were found in organic milk in the winter compared to the other systems. The literature shows that fat is the milk component that can present the greatest variation, and diet is responsible for 50% of this variation [37]. It is important to note that the current study used cows in similar lactation stages in the three studied management systems, and that O-LI cows received very little concentrate (Table 2). Therefore, the greatest levels of fat and total solids present in organic milk in the winter are likely due to pasture quality and diet, as seen in previous studies [38,39]. Despite differences in nutritional management shown in Table 2 and Table 4, no other difference was found in the quality of milk. Levels of protein and lactose in the milk averaged 3.28% ± 0.08% and 4.41% ± 0.05%, respectively, and did not vary between management types or across seasons (Table 3).

In relation to seasonality, the level of fat was lower in summer only for C-HI and higher in the winter for O-LI. We found no further significant differences. The lower fat for C-HI could be considered unexpected due to the higher fiber content in pasture occurring in this season [40]. C-HI farms had a higher proportion of crops that occur mostly in the summer. Therefore, in the summer we can expect a lower pasture area for these farms, resulting in an increased speed of rotation on their pasture, composed only of highly digestible annual species (Table 4). It was not possible to analyze the bromatological composition of the diet. Therefore, we can only infer that the summer pastures were grazed too early, when pasture composition was likely very low in fiber. Regarding the higher fat content in the milk from O-LI farms in the winter, we can only infer that the opposite occurred. Although the cows were fed ryegrass, oats and clover, three highly digestible species, there was also a high prevalence of tropical species in the diet, which would provide high levels of fiber. In fact, in our analysis we observed the cows eating these species (Table 4).

Considering the hygienic quality of milk, no significant differences were found between management types and seasons for BMBC or BTSCC (Table 5). Overall, the values found for BTSCC are within the limits allowed by Brazilian legislation, but the maximum values for BMBC found are alarming. At the time of the study, Brazilian legislation allowed a maximum value for both BTSCC and BMBC of 600,000 per mL, and the results demonstrate the inability of many farms to comply with existing regulations. Regarding milking conditions in the farms studied, all used milking machines, but on the majority of farms (15 out of 24) the milk was collected in a receiving jar and manually transferred to a cooling tank. On five farms, the milk collected was kept in the jars, which was placed inside cool water in cooling tanks. Only on five farms—four C-HI and one C-LI—was a pipeline system in place to transport the milk to the cooling tank. Excessive manipulation of the milk, such as the manual transfer from one recipient to another, and delay in the transfer to the cooling tank may influence contamination and growth of bacteria [41]. The dairy company collected milk on average every second day (range 1–4 days), which could influence high BMBC in general. Nonetheless, we found no differences among management types. We could not identify any influence of the type of equipment or milking parlor environment on BMBC data. Because of the paramount importance of this information for human health, further studies must be done to identify the causes of high BMBC in order to develop strategies to mitigate against such high bacteria counts.

The main factors contributing to increased BMBC in milk are post-harvest and related to environment and hygiene: ambient temperature, accumulation of mud due to periods of rain resulting in unclean teats, contaminated facilities, improper management and poor cleaning of milking machines, inadequate refrigeration of milk, and contact of milk with air, among others [42]. Moreover, bacterial contamination of raw milk can generally occur from three main sources: within the udder, outside the udder (cow environment) and the surface of milk handling and storage equipment. The degree of cleanliness of the milking system is a major factor influencing the total bulk milk bacteria count [41]. Considering the low risk of the cows' environment in the present study (cows spent most of the time on pasture), the high BMBC is probably due to a lack of cleaning and sanitizing procedures in milking machines and storage equipment and the high BTSCC due to high levels of mastitis.

Comparing the standards established by the principal milk-producing countries worldwide for BTSCC and BMBC, we can see some differences among them. The majority of European countries, New Zealand and Australia have maximum limits of BTSCC of 400,000, in Canada the limit is 500,000 and in the USA the limit is 750,000 cells/mL [43]. In the USA, there have been attempts to reduce this limit to 400,000 cells/mL, but the National Conference on Interstate Milk Shipments [44] did not find a hygiene-related justification for such a modification. The maximum BMBC permitted for all countries noted above is 100,000 CFU/mL. Regardless of the management type used, we found higher values than those permitted by the Brazilian legislation at the time of the study [24] on a significant number of farms primarily for BMBC. As of July 2016, Brazilian law will require these values to be below 400,000 cells/mL for BTSCC and 100,000 CFU/mL for BMBC, which presents a major challenge for the dairy industry. In order to adapt to these new regulations, farmers must adopt drastic measures to control mastitis and hygiene during and after milking.

Table 5. Bulk Milk Bacteria Count (BMBC)* and Bulk Tank Somatic Cell Count (BTSCC) in milk collected from cooling tanks on the farms grouped by conventional high input (C-HI), conventional low input (C-LI), and organic low input (O-LI) (n = 8 per group) management (n = 8 per group).

Season	Management Types	BMBC (× 1000 CFU/mL)	SEM	BTSCC (× 1000 cells/mL)	SEM
Autumn	C-HI	1226	724.4	775	231.7
	C-LI	1408	724.4	545	231.7
	O-LI	1472	724.4	609	231.7
Winter	C-HI	223	958.4	498	113.4
	C-LI	2174	958.4	469	113.4
	O-LI	498	958.4	350	113.4
Spring	C-HI	418	197.4	301	85.6
	C-LI	475	197.4	575	85.6
	O-LI	801	197.4	297	85.6
Summer	C-HI	444	131.6	353	97.4
	C-LI	557	131.6	467	97.4
	O-LI	393	131.6	308	97.4

* Mean values represented. SEM = standard error of the mean.

From the distribution of the studied variables in the total data set matrix, we can see a clear separation between organic farms and the conventional farms in axis PC1, which explains 76% of the variability in the data (Figure 1). The O-LI farms are located in PC1+ and the conventional farms in PC1-, with no distinction between farms managed by C-HI or C-LI (Figure 1). The descriptors that contribute to a classification of samples in PC1 and PC2 are the percentages of total farm area designated for perennial and annual pasture, respectively (Figure 2). In this case, the land use was the factor responsible for the grouping (Figure 2). While other attributes initially considered important for the distinction between groups were included in the analysis, such as the use of supplements in cow diet, the results showed little significance. In previous studies, PCA was able to demonstrate a distinction of milk samples based on the region or management system used [11,45].

Above, we mentioned the results of the PCA analysis when noting that on average half of the area of organic farms was designated for the cultivation of perennial pasture, representing almost double that used in conventional management (Table 1). The extent of this finding goes beyond the specific differences in land use, given that the perennial vegetation more effectively provides essential benefits for eco-efficiency of the management system. Previous studies show that annual crops result in significant reductions in soil fertility and require intensive inputs to maintain productivity, while perennial pastures maintain high levels of soil fertility as well as more complex biological communities [31,46]. Furthermore, in terms of the trophic chain, perennial pasture promotes a greater number and diversity of pollinating insects, herbivores and decomposers [47]. As such, perennial crops are a valuable ecological resource for sustainable agriculture.

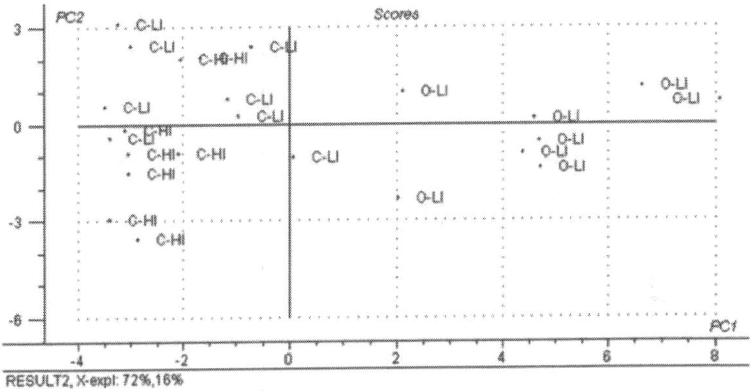

Figure 1. Factorial distribution of principal component 1 (PC1) and principal component 2 (PC2) for all studied descriptors of farms grouped by conventional high input (C-HI), conventional low input (C-LI) and organic low input (O-LI) (n = 8 per group).

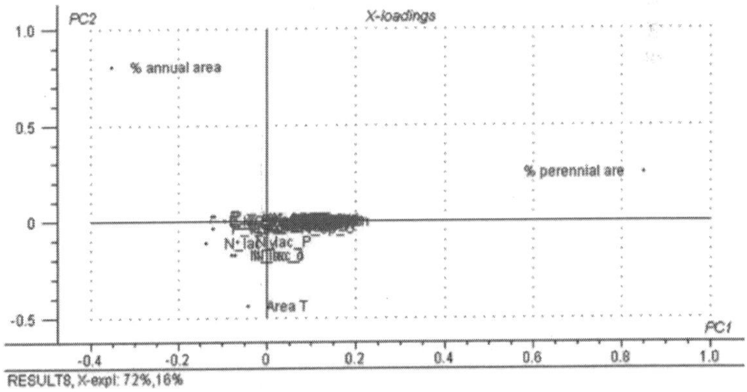

Figure 2. Factorial contribution of principal component 1 (PC1) and principal component 2 (PC2) for the classification of farms based on all descriptors: total farm area, % of area annual pasture, % of area perennial pasture, average number of forage species, number of species per season, intake of concentrate and silage per day and cow in each season, % of fat, protein, lactose, total solid, non-fat dry extract, BTSCC and BMBC values.

4. CONCLUSIONS

The three pasture-based systems differed in type of pasture (perennial or annual), farm size, intake of concentrate and silage and pasture diet. The O-LI had the smallest farms and herds, lowest input and lowest milk yields compared to C-HI and C-LI. C-HI had higher inputs and milk yields than C-LI. Total milk composition was little affected by production system and season, but BTSCC was generally high and BMBC exceeded the acceptable level in all systems and all seasons. Based on the available data we may conclude that C-HI and O-LI systems are economically more beneficial for the farmer; however, the low milk yields on O-LI farms are a major concern. Besides milk production, O-LI farms may contribute to important eco-system services. Cell count and hygienic quality need to be addressed in all production systems especially during the autumn.

ACKNOWLEDGMENTS

We would like to acknowledge CNPq for financial support of the research through the Inter-institutional Chain of Production Network for Agroecological Milk (Rede Interinstitucional da Cadeia Produtiva do Leite Agroecológico), nº 562908/2010-2, Edital 22/2010—Theme D. We also thank Evelyn Nimmo for the editing of the manuscript.

AUTHOR CONTRIBUTIONS

Conceived and designed the study: SK, RBS, LCPMF, LAH. Data collection: RBS. Analyzed the data: LAH, LCPMF, SK. Contributed reagents/materials/analysis tools: LCPMF, SK. Wrote the paper: SK, RBS, LAH, LCPMF.

CONFLICTS OF INTEREST

The authors declare no conflict of interest.

REFERENCES

1. FAO. Food and Agriculture Organization of the United Nations—Statistics Division. 2012. Available online: http://faostat.fao.org/site/339/default.aspx (accessed on 20 January 2015).

2. Fischer, A.; Santos Junior, S.; Sehnem, S.; Bernardi, I. Produção e produtividade de leite do oeste catarinense. RACE-Revista de Administração, Contabilidade e Economia 2012, 10, 337–362. (in Portuguese).

3. ICEPA. Sintese Anual da Agricultura de Santa Catarina. 2009. Available online: http://www.cepa.epagri.sc.gov.br/Publicacoes/SINTESE2001.PDF (accessed on 15 July 2014). (in Portuguese).

4. Costa, J.H.C.; Hötzel, M.J.; Longo, C.; Balcão, L.F. A survey of management practices that influence production and welfare of dairy cattle on family farms in southern Brazil. J. Dairy Sci. 2013, 96, 307–317.

5. European Commission. Organic Farming. Available online: http://ec.europa.eu/agriculture/organic/organic-farming(accessed on 21 August 2013).

6. WHO. Global Burden of Disease in 2002: Data Sources, Methods and Results; WHO: Geneva, Switzerland, 2003; p. 181.

7. Honorato, L.A.; Machado Filho, L.C.P.; Barbosa Silveira, I.D.; Hötzel, M.J. Strategies used by dairy family farmers in the south of Brazil to comply with organic regulations. J. Dairy Sci. 2014, 97, 1319–1327.

8. Basset-Mens, C.; Ledgard, S.; Boyes, M. Eco-efficiency of intensification scenarios for milk production in New Zealand. J. Ecol. Econ. 2009, 68, 1615–1625.

9. Thomassen, M.A.; Van Calker, K.J.; Smits, M.C.J.; Iepema, G.L.; De Boer, I.J.M. Life cycle assessment of conventional and organic milk production in the Netherlands. Agric. Syst. 2008, 96, 95–107.

10. Butler, G.; Nielsen, J.H.; Slots, T.; Seal, C.; Eyre, M.D.; Sanderson, R.; Leifert, C. Fatty acid and fat-soluble antioxidant concentrations in milk from high- and low-input conventional and organic systems: Seasonal variation. J. Sci. Food Agric. 2008, 88, 1431–1441.

11. Slots, T.; Butler, G.; Leifert, C.; Kristensen, T.; Skibsted, L.H.; Nielsen, J.H. Potentials to differentiate milk composition by different feeding strategies. J. Dairy Sci. 2009, 92, 2057–2066.

12. Stiglbauer, K.E.; Cicconi-Hogan, K.M.; Richert, R.; Schukken, Y.H.; Ruegg, P.L.; Gamroth, M. Assessment of herd management on organic and conventional dairy farms in the United States. J. Dairy Sci. 2013, 96, 1290–1300.

13. Adler, S.; Jensen, S.K.; Govasmark, E.; Steinshamn, H. Effect of short-term versus long-term grassland management and seasonal variation in organic and conventional dairy farming on the composition of bulk tank milk. J. Dairy Sci.2013, 96, 5793–5810.

14. Pulido, R.G.; Muñoz, R.; Lemarie, P.; Wittwer, F.; Orellana, P.; Waghorn, G.C. Impact of increasing grain feeding frequency on production of dairy cows grazing pasture. Livest. Sci. 2009, 125, 109–114.

15. Soder, K.J.; Rotz, C.A. Impact of Utilizing a Total Mixed Ration in Pennsylvania Grazing Dairy Herds; USDA/Agricultural Research Service, Pasture Systems and Watershed Management Research Unit: University Park, PA, USA, 2007.

16. Balcão, F.L. Tipologia da atividade leiteira na região Noroeste de Santa Catarina. Master Degree, Universidade Federal de Santa Catarina, Florianópolis, Brazil, 2012.

17. Brasil. Lei n.10.831, de 23 de dezembro de 2003. Dispõe sobre a agricultura orgânica. IBD Certificações. Diretrizes e Legislação. Decreto da Lei 10.831 de Produtos Orgânicos, Available online: http://www.ibd.com.br/Downloads/DirLeg/Legislacao/05-Lei_10831_2003.pdf (accessed on 13 September 2014).

18. Alvares, C.A.; Stape, J.L.; Sentelhas, P.C.; de Moraes Gonçalves, J.L.; Sparovek, G. Köppen's climate classification map for Brazil. Meteorologische Zeitschrift 2014, 22, 711–728.

19. Van Soest, P.J.; Goering, H.K. Forage Fiber Analysis: Apparatus, Reagents, Procedures and Some Applications; Agriculture Research Service, United States Department of Agriculture: Washington, DC, USA, 1970; pp. 1–20.

20. Euclides, V.P.B.; Macedo, M.C.M.; Oliveira, M.P. Avaliação de diferentes métodos de amostragens sob pastejo. R. Bras. Zootec. 1992, 21, 691–702.

21. IDF. International IDF Standard; 141C; IDF: Brussels, Belgium, 2000.

22. IDF. International IDF Standard; 196; IDF: Brussels, Belgium, 2004.

23. IDF. International IDF Standard; 148-2; IDF: Brussels, Belgium, 2006.

24. Brasil. Instrução normativa n.46, de 6 de outubro de 2011. Ministério da Agricultura, Pecuária e Abastecimento, 2011. Available online: http://www.aao.org.br/aao/pdfs/legislacao-dos.../instrucao-normativa-n46.pdf (accessed on 21 August 2013).

25. Clive, J. Global Status of Commercialized Biotech/GM Crops: 2013; ISAAA Brief N° 46. ISAAA: Ithaca, NY, USA, 2013. Available online: http://www.isaaa.org/resources/publications/briefs/46/executivesummary/ (accessed on 9 April 2015).

26. Voisin, A. Produtividade do Pasto; Mestre Jou: São Paulo, Brazil, 1974; p. 520.

27. Machado, L.C.P. Pastoreio Racional Voisin; Cinco Continentes: Porto Alegre, Brazil, 2004.

28. Hart, R.H.; Bissio, J.; Samuel, M.J.; Waggoner, J.W., Jr. Grazing systems, pasture size, and cattle grazing behavior and gains. J. Range Manag. 1993, 46, 81–87.

29. Abrahamse, P.A.; Dijkstra, J.; Vlaeminck, B.; Tamminga, S. Frequent allocation of rotationally grazed dairy cows changes grazing behavior and improves productivity. J. Dairy Sci. 2008, 91, 2033–2045.

30. Pavlu, V.; Hejcmanb, M.; Gaisler, J.; Nezerková, P. Effect of continuous grazing on forage quality, quantity and animal performance. Agric. Ecosyst. Environ. 2006, 113, 349–355.

31. Tilman, D.; Reich, P.B.; Knops, J.; Wedin, D.; Mielke, T.; Lehman, C. Diversity and Productivity in a Long-Term Grassland Experiment. Science 2001, 294, 843–845.

32. Finlayson, J.D.; Lawes, R.A.; Metcalf, T.; Robertson, M.J.; Ferris, D.; Ewing, M.A. A bio-economic evaluation of the profitability of adopting subtropical grasses and pasture-cropping on crop-livestock farms. Agric. Syst. 2012, 106, 102–112.

33. Sanderson, M.A. Stability of production and plant species diversity in managed grasslands: A retrospective study.Basic Appl. Ecol. 2010, 11, 216–224.

34. Sanderson, M.A.; Corson, M.S.; Rotz, C.A.; Soder, K.J. Economic Analysis of Forage Mixture Productivity in Pastures Grazed by Dairy Cattle. Online. Forage Grazinglands 2006, 4. Available online: https://www.crops.org/publications/fg/abstracts/4/1/2006-0929-01-RS?access=0&view=article (accessed on 13 July 2013).

35. Nauta, W.J.; Baars, T.; Saatkamp, H.; Weenink, D.; Roep, D. Farming strategies in organic dairy farming: Effects on breeding goal and choice of breed. An explorative study. Livest. Sci. 2009, 121, 187–199.

36. Clark, D.A.; Caradus, J.R.; Monaghan, R.M.; Sharp, P.; Thorrold, B.S. Issues and options for future dairy farming in New Zealand. N. Zeal. J. Agric. Res. 2007, 50, 203–221.

37. Fredeen, A.H. Considerations in the nutritional modification of milk composition. Anim. Feed Sci. Technol. 1996, 59, 185–197.

38. Teixeira, N.M.; Freitas, A.F.; Barra, R.B. Influência de fatores de meio ambiente na variação mensal da composição e contagem de células somáticas do leite em rebanhos no estado de Minas Gerais. Arq. Bras. Med. Vet. Zootec. 2003, 55, 491–499.

39. Lopes, F.C.F.; Aroeira, L.J.M.; Rodriguez, N.M.; Deresz, F.; Sampaio, I.B.M.; Paciullo, D.S.C.; Vittori, A. Efeito da suplementação e do intervalo de pastejo sobre a qualidade da forragem e consumo voluntário de vacas Holandês x Zebu em lactação em pastagem de capim elefante. Arq. Bras. Med. Vet. Zootec. 2004, 56, 355–362.

40. Aroeira, L.J.M.; Lopes, F.C.F.; Deresz, F.; Verneque, R.S.; Dayrell, M.S.; De Matos, L.L.; Maldonado-Vasquez, H.; Vittori, A. Pasture availability and dry matter intake of lactating crossbred cows grazing elephant grass (Pennisetum purpureum, Schum.). Anim. Feed Sci. Technol. 1999, 78, 313–324.

41. Law, B.A.; Tamime, A.Y. Technology of Cheesemaking, 2nd ed.; Wiley-Brackwell Publishing: Ames, IA, USA, 2010.

42. Bueno, V.F.F.; De Mesquita, A.J.; de Oliveira, A.N.; Nicolau, E.S.; Neves, R.B.S. Contagem bacteriana total do leite: Relação com a composição centesimal e período do ano no Estado de Goiás. Revista Brasileira de Ciência Veterinária2008, 15, 40–44.

43. APHIS. Determining U.S. Milk Quality Using Bulk-tank Somatic Cell Counts. Veterinary Services, Centers for Epidemiology and Animal Health, 2012. Available online: http://usdasearch.usda.gov/search?utf8=%3F&affiliate=usdaaphis&query=somatic+cell&commit=Search (accessed on 17 March 2015).

44. NCIMS—National Conference on Interstate Milk Shipments. Actions of the 2011 National Conference on Interstate Milk Shipments. Available online: http://www.ncims.org/ (accessed on 28 May 2013).

45. Larsen, M.K.; Nielsen, J.H.; Butler, G.; Leifert, C.; Slots, T.; Kristiansen, G.H.; Gustafsson, A.H. Milk quality as affected by feeding

regimens in a country with climatic variation. J. Dairy Sci. 2010, 93, 2863–2873.

46. Culman, S.W.; Dupont, S.T.; Glover, J.D.; Buckley, D.H.; Fick, G.W.; Ferris, H.; Crews, T.E. Long-term impacts of high-input annual cropping and unfertilized perennial grass production on soil properties and belowground food webs in Kansas, USA. Agric. Ecosyst. Environ. 2010, 137, 13–24.

47. Glover, J.D.; Culman, S.W.; Dupont, S.T.; Broussard, W.; Young, L.; Mangan, M.E.; Mai, J.G.; Crews, T.E.; Dehaan, L.R.; Buckley, D.H.; et al. Harvested perennial grasslands provide ecological benchmarks for agricultural sustainability. Agric. Ecosyst. Environ. 2010, 137, 3–12.

CHAPTER 9

Compost Barns for Dairy Cows—Aspects of Animal Welfare

E. Ofner-Schröck[1*], M. Zähner[2], G. Huber[1], K. Guldimann[2], T. Guggenberger[1], J. Gasteiner[1]

[1]HBLFA Raumberg-Gumpenstein, Agricultural Research & Education Centre, Irdning, Austria

[2]Institute for Sustainability Sciences (ISS), Ettenhausen, Switzerland

ABSTRACT

Compost barns for dairy cows are showing increased popularity also in Central Europe. A compost barn is used mainly as a two-area system with a bedded lying area and a solid feeding alley. Sawdust or dry fine wood shavings or wood chips are mostly used as bedding material, which has to be stirred twice a day. Stirring aerates and mixes faeces and urine into the bedding material, the mixture decomposes by means of aerobic microorganisms. A joint research project between the Agricultural Research and Education Centre Raumberg-Gumpenstein (HBLFA) and the Institute for Sustainability Sciences Tänikon (ISS) analyzed amongst other things, the cleanliness of the animals, integument alterations, lying behaviour and the current lameness situation of animals. A total of 138 cows were examined on five Austrian dairy farms. All cows were visually scored and animal behaviour was observed by data loggers as well as by direct observation. The mean value concerning cleanliness of animals was 0.44, while the udder was the cleanest and the lower leg the dirtiest area. Only a few lesions in carpal and tarsal joints could be found. Cows showed no differences in lying behaviour between times of day and temperatures. Large differences in lying behaviour were evident among farms. While on the compost barn farms only around 25% of all cows were scored to be lame, on cubicle-housing system farms 31% - 46% of the cows fell into that category ($p < 0.001$). From the present results, the compost barn can be seen as an animal-friendly system. In further investigations other factors affecting animal health and to resolve any outstanding issues concerning economy and alternative litter materials should be analyzed.

KEYWORDS

Compost Barn, Cattle, Dairy Cows, Behaviour, Cleanliness, Lameness

1. INTRODUCTION

Compost barns for dairy cows are also becoming more and more popular in Central Europe. They have already been successfully built and operated for quite some time in Israel and America. A lot of the scientific findings currently available on this kind of housing system come from America, especially Minnesota [1] -[5] .

A compost barn is used mainly as a two-area system with a bedded lying area and a feeding alley, which can have either a solid or a slatted floor. Regarding bedding, good experience has been gained in Austria with sawdust and wood shavings. Starting with a bed approximately 20 to 25 cm high (1.8 - 2.5 m^3/animal), about 0.4 to 1.3 m^3/animal is added every 2 to 7 weeks (10 - 15 m^3/animal and year). The bedding material is loosened once or twice a day with a grubber or rotary hoe to a depth of 20 to 25 cm, working in any faeces and urine. This lets air into the bedding so that the mixture can decompose with the help of aerobic micro-organisms [6] . The result is a loose, friable bedding material. Its temperature should be between 30˚C and 45˚C in order to quickly convert the organic substance and kill off undesirable germs while supporting beneficial germs [7] . As far as possible, new bedding should not be started during the winter months because it is difficult for the rotting process to begin in the cold. The lying area may be separated from the feeding alley by walls or it may be placed about 30 to 50 cm lower. Compost removal (mucking out) is performed twice a year (in spring and autumn) when the compost bedding has reached a thickness of 50 to 60 cm. The size of the lying area is of crucial importance not only for the cleanliness and wellbeing of the animals but also for the efficiency of the housing system. While in Israel there is an area of between 13 and 20 m^2 for each animal, in the USA the lying areas are 7.5 to 9.2 m^2 per cow [8] . With regard to the size of lying area, the division of space and the management (bedding, and dung removal), the systems commonly used in Austria are more similar to the American system. Positive effects on efficiency are seen with simple structures with low construction costs and a relatively small amount of working hours needed to care for the lying areas. However, the availability of cheap bedding materials is of great importance with regard to the efficiency of the compost housing system. Economical sources of supply have to be found on individual farms, and further research work has to be performed on alternative bedding materials. In any case, the compost barn can be considered as an interesting system for grassland farms without a great deal of straw available.

2. MATERIALS AND METHODS

A joint research project between the Agricultural Research and Education Centre Raumberg-Gumpenstein (HBLFA) and the Institute for Sustainability Sciences Tänikon (ISS) analyzed the overall conditions involved in the construction and running of compost barns for dairy farming. Amongst other things, the integument alterations, cleanliness of the animals, lying behaviour and the current lameness situation of animals were examined.

Investigations into animal behaviour, cleanliness of the animals, integument alterations and lameness assessments were performed on five Austrian farms with compost barns in Upper Austria and Styria. A total of 138 cows were examined. The herd size on the farms varied between 18 and 35 cows. Another two farms in Upper Austria were included in a questionnaire survey, so a total of seven farms with compost barns.

2.1. Animal Behaviour

The place selected by the animals for lying and standing was investigated by direct observation. On each farm, on two days, the behaviour of 10 focus animals was recorded during the periods from 10:30 to 12:30 hrs and from 14:00 to 16:00 hrs, in 5-minute intervals. It was noted whether and where the animals were lying—on the edge (near the outside wall, near the inside wall), in the centre or outside of the lying area. The places found for standing at the particular time was also documented.

2.2. Cleanliness of the Animals

The cleanliness of the animals was assessed using the model devised by Faye and Barnouin [9] on all of the animals in the herd. With this model, the animals are assessed on five areas of the body (zones) with five scores (0 = no dirt to 2 = totally dirty or covered with thick plaques of dirt):

Zone 1: Area between the base of the tail, the ischial tuberosity and the area of the suspensory apparatus of udder;

Zone 2: Udder viewed from behind;

Zone 3: Lower hind leg (area of the tarsus to the dewclaw);

Zone 4: Udder viewed from the side;

Zone 5: Upper hind leg (excluding the tarsus).

2.3. Integument Alterations (Hairless Patches, Lesions, Swellings)

The integument alterations were also assessed with all of the animals in the herd, using the system devised by Ekesbo [10] . The assessment was limited to the carpus, tarsus and the tarsal joint. The examination included hairless patches, dry scabs and open wounds (according to size) and swellings.

2.4. Lameness

In this project the animals were assessed according to a model developed by Winckler and Willen [11] , in which the degree of lameness was classified according to five scores (values):

Score 1: Normal gait;

Score 2: Uneven gait;

Score 3: Short striding gait with one limb;

Score 4: Short striding gait with more than one limb or strong reluctance to bear weight on one limb;

Score 5: Does not support on one limb or strong reluctance to put weight on limb in two or more limbs.

Before being assessed, the cows were locked in the feeding rack and then released individually and assessed while walking in the feeding alley.

2.5. Statistical Analysis

The data were analyzed with Microsoft Excel and the statistics software packages R and Statgraphics Centurion.

3. RESULTS AND DISCUSSION

This article gives selected results from the research project entitled "Overall conditions for the use of compost barns in dairy farming" ("Rahmenbedingungenfür den Einsatz von Kompostställen in der Milchviehhaltung"). A full indication of all of the results is given in the final report on that project [12].

3.1. Animal Behaviour

The proportions of cows standing and lying show that, in addition to the differences between morning and afternoon, the great differences are mainly between farms (Table 1). Within the present temperature range it was not possible to observe any connection between the temperature and the proportion of cows lying. Even at higher temperatures no avoidance of the lying areas could be discerned. Furthermore, the lying area was very seldom used for standing and only by a few animals. Regarding the location selected for lying, there were no differences between morning and afternoon (Figure 1). Here the differences between farms were much greater, too. The choice of lying locations and the overall lying behaviour were greatly influenced by the structural arrangement and the shape of the lying area (long and narrow to short and wide). The open construction with good ventilation in all farms allowed for a good stable climate even in high temperatures. Zähner [13] was able to show in investigations that cows prefer open lying areas in summer and more protected areas in winter.

3.2. Cleanliness of the Animals

Table 2 summarizes the results on cleanliness of the animal. The mean value across all zones was 0.44. The most dirty area of the body was zone 3 (lower hind leg) with an average score of 0.80; the least dirtiness was in zone 4 (udder seen from the side) with an average score of 0.19. As shown in Figure 2, the cleanliness of the animals in compost barns is comparable with other housing systems [14] [15] . Hörning [16] found an average dirtiness of 0.40 (54 farms) in cubicle housing systems, 0.59 (30 farms) in deep-litter systems and 0.77 (29 farms) in sloped-floor systems, but stressed the high levels of variation within one system.

The animals in the compost barn display a good level of cleanliness. The greater degree of dirtiness in the area of the lower leg is probably incurred mostly in the feeding alleys. The very low level of dirtiness on the udder is a positive observation with regard to udder health, because a dirty udder is in particular connected with a high risk of infection with pathogens [17] . Influences specific to individual farms, such as stocking density, as well as the management of the compost barn for instance the frequency of providing fresh bedding and dung removal (compost removal), bedding material and amount of bedding as well as the "consistency" of the animals' faeces should be taken into account when interpreting the dirtiness of the animals.

Table 1. Proportion of standing and lying cows and temperature in the two-day period on the five farms with compost barns examined.

Period	Farm	Proportion of standing animals [%]			Proportion of lying animals [%]			Temperature [°C]		
		Average	Min.	Max.	Average	Min.	Max.	Average	Min.	Max.
Morning	1	76.4	46.4	100.0	23.6	0.0	53.6	16.5	16.1	16.9
	2	45.3	13.0	100.0	54.7	0.0	87.0	15.4	12.9	18.5
	3	51.8	31.3	78.8	48.2	21.2	68.8	19.8	17.0	22.4
	4	42.8	10.0	90.0	57.2	10.0	90.0	20.9	19.6	25.0
	5	59.8	10.0	100.0	40.2	0.0	90.0	20.4	19.1	26.0
Afternoon	1	36.5	10.7	64.3	63.5	35.7	89.3	19.0	17.5	21.3
	2	22.6	0.0	100.0	77.4	0.0	100.0	17.6	13.9	22.2
	3	69.6	33.3	100.0	30.4	0.0	66.7	21.1	18.8	23.9
	4	55.5	30.0	80.0	44.5	20.0	70.0	22.5	21.2	24.0
	5	71.5	40.0	100.0	28.5	0.0	60.0	21.7	20.7	22.6

Table 2. Dirtiness of the animals broken down according to farms and zones (areas of the body).

Farm	Number of animals	Dirtiness (mean values)						Total for zones (index)
		Zone					Mean value for all zones	
		Base of tail	Rear of udder	Lower leg	Side of udder	Upper leg		
1	27	0.70	0.26	1.26	0.26	0.81	0.66	3.29
2	23	0.72	0.41	1.04	0.30	0.37	0.57	2.84
3	32	0.50	0.25	0.56	0.14	0.14	0.32	1.59
4	18	0.21	0.16	0.26	0.08	0.21	0.18	0.92
5	35	0.51	0.29	0.79	0.19	0.41	0.44	2.19
Mean value for all cows	n = 135	0.54	0.28	0.80	0.19	0.39	0.44	2.17

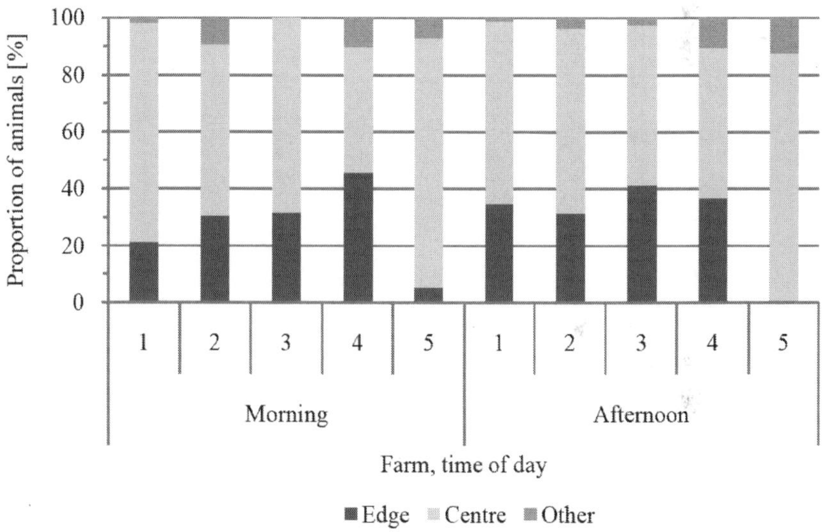

Figure 1. Proportion of cows lying in different areas of the lying area—edge, centre, and others.

3.3. Integument Alterations

In the farms with compost barns investigated, there were only slight integument alterations in the animals (n = 135). Hairless patches < 2 cm appear on average in 9.6% of all cows (Max.: 25.9%; Min.: 2.2%). Hairless patches > 2 cm could be seen on average in 2.2% of all cows (Max.: 12.6%; Min.: 0%). Dry scabs < 2 cm appear on average in 0.7% of all cows (Max.: 4.4%; Min.: 0%). There was swelling on only one cow and no open wounds could be detected. Comparing the proportion of animals with an intact hair coat (no integument alteration) on the joints with values from other housing systems gives the image shown in Table 3. The extent of various integument alterations in different housing systems ($p < 0.001$) is shown in Figure 3.

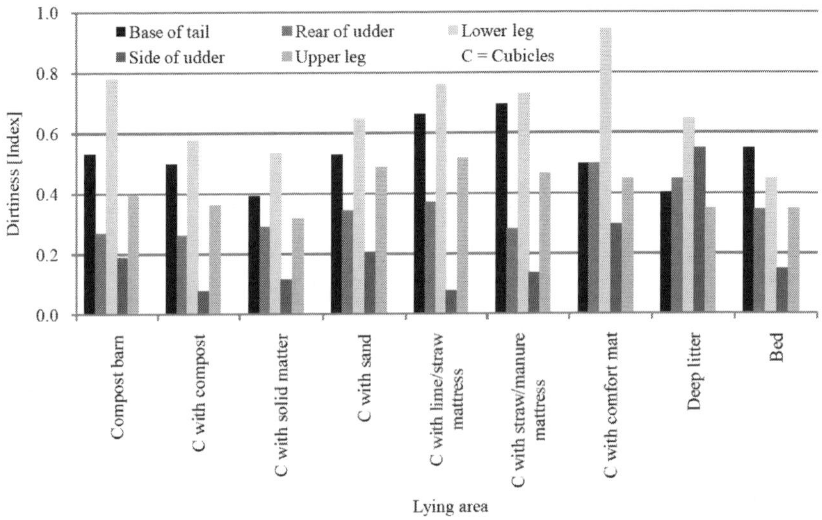

Figure 2. Dirtiness of the animals in different housing systems (comparative values on compost barns from [14] [15]).

In the compost barn an average of 82.6% of all of the animals displayed an intact hair coat (no integument alteration) on the joints. According to the aforementioned authors, this is a value that is comparable with a cubicle housing system with straw-manure mattress (86.2%). The investigation performed by Barberg et al. [1] on animal welfare in compost barns in Minnesota (USA) there were alterations to the tarsus in 25% of the animals (n = 796). In that case, 24% of the animals had hairless patches and the remaining 1% of the animals had swellings. No alterations were found in seven out of twelve herds.

3.4. Lameness

If we consider all of the cows observed (n = 138), 42.0% display a score of 1; 31.9% a score of 2; 16.7% a score of 3; 6.5% a score of 4 and 2.2% a score of 5 (Figure 4). In further analysis, to have a better overview and be able to compare the values with the data in the literature, scores 3 + 4 + 5 were classed together as “lame” and 1 + 2 as “not lame” (Table 4). In all of the cows observed (n = 138), 73.9% could be classified as “not lame” and 25.4% as “lame”. These are compared with values from lameness assessments that were performed using exactly the same methodology in ten cubicle housing systems in Styria [20] . In all of the cows observed in that case (n = 175), 54.3% could be classified as “not lame” and 45.7% as “lame”. Figure 4 shows the degree of lameness found on farms with cubicle housing systems.

If we compare the results of the assessment on the five compost barn farms and the ten farms with cubicle housing systems, significant differences appear in the number of lame cows ($p < 0.001$). While on the compost barn farms only around 25% of all cows were classed as “lame”, on the cubicle-housing system farms around 46% fell into that category. Continuing investigations to analyze other influencing factors (e.g. feeding, claw trimming) and to increase the sample should be sought. Investigations by other authors on Austrian farms with cubicle housing systems showed a proportion of 31% lame animals (sample size = 30 farms [21]) and 36% lame animals (sample size = 80 farms [22]). In the investigation performed by Barberg et al. [1] in compost barns in Minnesota (USA), 7.8% of the animals (n = 793) displayed clinical lameness using the five-stage system established by Sprecher et al. [23] . There were no lame cows to be found in two out of twelve herds.

Table 3. Proportion of animals with an intact hair coat (no integument alteration) on the joints in different housing systems [%] (comparative values on compost barn from [15] [18] [19]).

	Mean value	Standard deviation	Maximum	Minimum
Compost barn	82.6	18.6	97.1	52.2
Cubicles with compost	59.3	10.4	66.7	52.0
Cubicles with solid matter	59.3	13.1	72.3	462
Cubicles with sand	52.3	11.4	65.0	37.5
Cubicles with lime-straw mattress	78.2	9.8	82.0	58.0
Cubicles with straw-manure mattress	86.2	10.6	95.0	68.2
Cubicles with loose straw	32.1	33.9	76.5	0.0
Cubicles with comfort mat	14.6	14.3	44.4	0.0
Cubicles with rubber mat	9.8	6.2	18.8	4.8

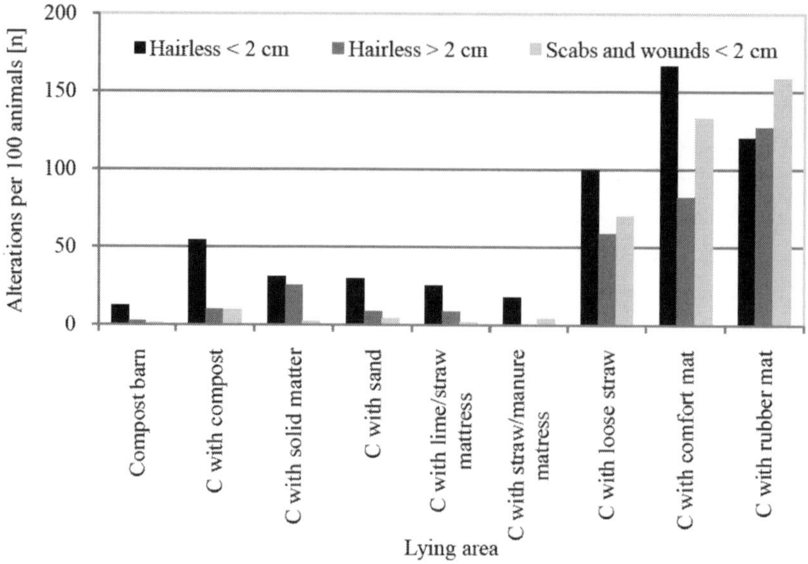

Figure 3. Animals with alterations to the joints in different housing systems (comparative values on compost barn from [15] [18] [19]).

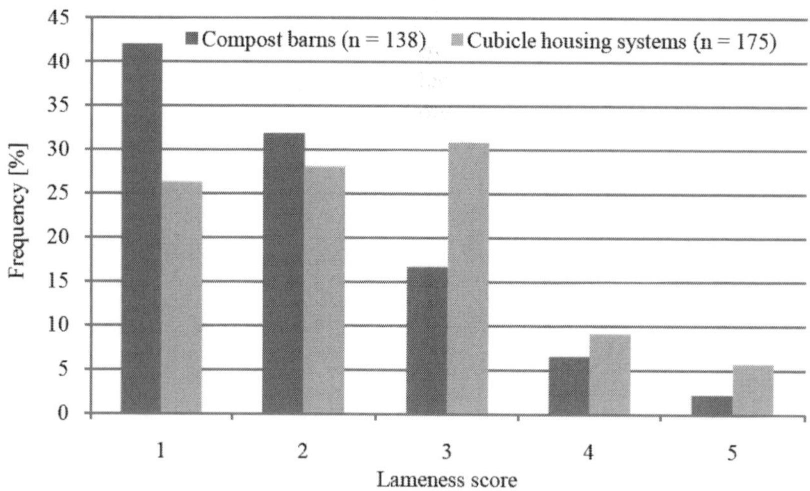

Figure 4. Degree of lameness in all of the cows (n = 138) on the farms with compost barns investigated compared to the degree of lameness in all of the cows (n = 175) in the farms with cubicle housing systems investigated according to Ofner-Schröck et al. [20] .

Table 4. Consolidated degree of lameness of cows kept in the compost barns (score 1 + 2 = "not lame"; score 3 + 4 + 5 = "lame").

Farm No.	Number of cows	% of "not lame" cows (score 1 + 2)	% of "lame" cows (score 3 + 4 + 5)
1	34	67.6	32.4
2	21	57.1	42.9
3	30	70.0	30.0
4	18	77.8	16.7
5	35	91.4	8.6
Total	138	73.9	25.4

4. CONCLUSION

From the present results, the compost barn can be seen as an animal-friendly system. Further investigations are desirable to analyze other factors affecting animal health and to resolve any outstanding issues concerning economy and alternative litter materials.

ACKNOWLEDGEMENTS

We would like to thank the farmers for their participation and support to this study. We thank the colleagues from the extension service for agricultural buildings of the Austrian provinces for the good cooperation. Additionally, we thank the employees of the HBLFA Raumberg-Gumpenstein and ISS Tänikon for their input in data collection and analysis.

REFERENCES

1. Barberg, A.E., Endres, M.I., Salfer, J.A. and Reneau, J.K. (2007) Performance and Welfare of Dairy Cows in an Alternative Housing System in Minnesota. Journal of Dairy Science, 90, 1575-1583.
2. Barberg, A.E., Endres, M.I. and Janni, K.A. (2007) Compost Dairy Barns in Minnesota: A Descriptive Study. Applied Engineering in

Agriculture, American Society of Agricultural and Biological Engineers, 23, 231-238.

3. Endres, M.I. and Barberg, A.E. (2007) Behavior of Dairy Cows in an Alternative Bedded-Pack Housing System. Journal of Dairy Science, 90, 4192-4200.

4. Espejo, L.A., Endres, M.I. and Salfer, J.A. (2006) Prevalence of Lameness in High-Producing Holstein Cows Housed in Freestall Barns in Minnesota. Journal of Dairy Science, 89, 3052-3058.

5. Janni, K.A., Endres, M.I., Reneau, J.K. and Schoper, W.W. (2007) Compost Dairy Barn Layout and Management Re- commendations. Applied Engineering in Agriculture, American Society of Agricultural and Biological Engineers, 23, 97-102.

6. Holzeder, S. (2011) Kompoststall—eine Alternative stellt sich vor. In: Tagungsbandzur Bautagung Raumberg-Gum- penstein 2011, HBLFA Raumberg-Gumpenstein, Irdning, 5-6. http://www.raumberg-gumpenstein.at

7. Holzeder, S. (2012) Komfort zum Wohlfühlen. Elite, 3, 54-59.

8. Leifker, A. (2010) Grenzenlose Freiheit? Top Agrar, 4, R6-R10.

9. Faye, B. and Barnouin, J. (1985) Objectivation de la propreté des vaches laitières et des stabulations—L'indice de propreté. Bulletin Technique C.R.Z.V. Theix, INRA, 59, 61-67.

10. Ekesbo, I. (1984) Methoden der Beurteilung von Umwelteinflüssen auf Nutztiere unter besonderer Berücksichtigung der Tiergesundheit und des Tierschutzes. Wien. Tierärztliche Monatsschrift, 71, 186-190.

11. Winckler, C. and Willen, S. (2001) The Reliability and Repeatability of a Lameness Scoring System for Use as an Indicator of Welfare in Dairy Cattle. Acta Agriculturae Scandinavica, Section A—Animal Science, Suppl. 30, 103-107.

12. Ofner-Schröck, E., Zähner, M., Huber, G., Guldimann, K., Guggenberger, T. and Gasteiner, J. (2014) Rahmenbedin- gungen für den Einsatz von Kompostställen in der Milchviehhaltung. Final Report, HBLFA Raumberg-Gumpenstein, Irdning.http://www.raumberg-gumpenstein.at

13. Zähner, M. (2001) Beurteilung von Minimalställen für Milchvieh anhand ethologischer und physiologischer Parameter. Dissertation ETH, Zürich.

14. Keck, M., Zähner, M. and Hauser, R. (2004) Minimalställe für Milchkühe bewähren sich: Empfehlungen für die Planung und den Betrieb. FAT Report No. 620, Forschungsanstalt Agroscope Reckenholz-Tänikon ART.

15. Zähner, M., Schmidtko, J., Schrade, S., Sachaeren, W. and Otten, S. (2009) Alternative Einstreumaterialien in Liege- boxen. In: Tagungsband Bautagung Raumberg-Gumpenstein 2009, HBLFA Raumberg-Gumpenstein, 33-38. http://www.raumberg-gumpenstein.at

16. Hörning, B. (2003) Nutztierethologische Untersuchungen zur Liegeplatzqualität in Milchviehlaufstallsystemen unter besonderer Berücksichtigung eines epidemiologischen Ansatzes. Postdoctoral Thesis (Habilitationsschrift), Universität Kassel, Witzenhausen.

17. Schreiner, D.A. and Ruegg, P.L. (2003) Relationship between Udder and Leg Hygiene Scores and Subclinical Mastitis. Journal of Dairy Science, 86, 3460-3465.

18. Buchwalder, T. (1999) Einfluss der Liegeplatzqualität auf das Verhalten und die Schäden bei Milchkühen im Boxenlaufstall. Interner Schlussbericht, Tänikon.

19. Schaub, J., Friedli, K. and Wechsler, B. (1999) Weiche Liegematten für Milchvieh-Boxenlaufställe-Strohmatratzen und sechs Fabrikate von weichen Liegematten im Vergleich. FAT Reports No. 529, Tänikon.

20. Ofner-Schröck, E., Gasteiner, J., Guggenberger, J., Mösenbacher-Molterer, I., Häusler, J., Krimberger, B., Zainer, I., Zainer, J., Finotti, E. and Bachler, C. (2009) Vergleich der Bewertung der Tiergerechtheit von Rinderhaltungssystemen mit dem Tiergerechtheitsindex TGI 35 L und mit Hilfe von ethologischen und tiergesundheitlichen Parametern. Final Report Project No. BAL 03 2324, HBLFA Raumberg-Gumpenstein, Irdning.

21. Dippel, S., Dolezal, M., Brenninkmeyer, C., Brinkmann, J., March, S., Knierim, U. and Winckler, C. (2009) Risk Factors for Lameness in Cubicle Housed Austrian Simmental Dairy Cows. Preventive Veterinary Medicine, 90, 102-112.

22. Mülleder, C. and Waiblinger, S. (2004) Analyse der Einflussfaktoren auf Tiergerechtheit, Tiergesundheit und Leistung von Milchkühen im Boxenlaufstall auf konventionellen und biologischen Betrieben unter besonderer Berücksichtigung der Mensch-Tier-Beziehung. Endberichtzum Forschungsprojekt 1267, Eigenverlag Wien, 184 p.

23. Sprecher, D.J., Hostetler, D.E. and Kaneene, J.B. (1997) A Lameness Scoring System that Uses Posture and Gait to Predict Dairy Cattle Reproductive Performance. Theriogenology, 47, 1179-1187.

CHAPTER 10

High Prevalence of Subclinical Mastitis and Multidrug Resistant Staphylococcus aureus Are a Threat to Dairy Cattle Production in Kiboga District (Uganda)

Keneth Iceland Kasozi[1,2], John Bosco Tingiira[2], Patrick Vudriko[1*]

[1]Central Diagnostic Laboratory, Department of Pharmacy, Clinics and Comparative Medicine, College of Veterinary Medicine, Animal Resources and Biosecurity, Makerere University, Kampala, Uganda

[2]Kiboga District Regional Veterinary Laboratory, Department of Production and Marketing, Kiboga District Local Government, Kiboga, Uganda

ABSTRACT

Subclinical mastitis (SCM) is one of the major factors affecting the productivity of dairy cattle all over the world. This study established the burden of SCM and determined the potent antibacterial formulation for control of Staphylococcus aureus (SA) related SCM in selected dairy cattle farms in Kiboga district. A total of 124 dairy cattle from 12 farms were screened for SCM using California Mastitis Test (CMT) from Kiboga Town-Council, Kapeke and Lwamata sub-counties. The offending bacteria were cultured and the antibiogram of SA was carried out using antibacterial susceptibility by the modified Kirby-Bauer disc diffusion method. Additional qualitative data on the factors that predispose cows to SCM was obtained through questionnaires and observation of milking Practice. The prevalence of SCM in the three sub-counties was 87.9%. Over 70% of the dairy cattle screened for SCM had more than 2 udder quarters affected. The majority (90%) of SCM was caused by mixed bacterial infections: Coagulase negative staphylococci (64.4%) and SA (16.6%) being the most prevalent. All the farmers (100%) lacked knowledge on SCM, udder towels, teat dipping and drug cow therapy. Overall, 71.4% of SA isolated was multi-drug resistant. There was a high level of resistance against penicillin (100%), neomycin (85.7%) and tetracycline (71.4%). In contrast, all the above isolates were susceptible to Trimethoprim-Sulphamethazole. In conclusion, the high burden of SCM and emergence of multidrug resistant SA are one of the constraints to dairy production in Kiboga district. Therefore, sensitization of dairy farmers in Kiboga district on proper hygienic, appropriate milking techniques and dry cow therapy using potentiated sulfonamide intra-mammary preparations are highly recommended in SA associated SCM.

KEYWORDS

Antibacterial Susceptibility, Dairy Cattle, Subclinical Mastitis, Staphylococcus aureus

1. INTRODUCTION

Subclinical mastitis (SCM) continues to be a silent nightmare to the economic propensity of the dairy sector in Uganda. In fact, 75% of economic losses of mastitis is caused by SCM [1] [2] , thus contributing to the poor performance associated with the dairy sector in Uganda. As the disease takes toll on the dairy industry in Uganda, early detection and management remains elusive due to lack of diagnostic services at local government levels. This has been accentuated by the weak extension services and poor animal health seeking behavior of farmers. Moreover, recent studies [1] [3] have revealed that dairy farmers lacked knowledge on the ideal milking techniques, milking hygiene and SCM. The major contagious mastitis causative pathogens are Staphylococcus aureus (SA) and Streptococcus agalactiae (SAG), while the minor contagious pathogen is Corynebacterium bovis (CB) [1] . The major environmental mastitis causative pathogens are other Streptococci (OS) and Coliforms (CO), while

Coagulase negative Staphylococci (CNS) are minor environmental pathogens [1] [4] [5] .

SA is the fourth most common subclinical mastitis pathogen in Uganda [6] . SA is considered one of the most problematic mastitis pathogens due to its strong virulence factors [4] . Its α-haemolysis virulence factor was reported [2] to cause fatal gangrenous mastitis. The infection due to SA is easily spread within a herd, and cows get infected primarily during the milking process. The most commonly isolated CNS species in SCM are S. epidermidis, S. simulans, S. chomogenes, S. xylosus and S. haemolyticus; they are generally considered less virulent than SA thus causing milder forms of SCM [4] . Therefore, udder tissue damage due to CNS is limited, hence the prognosis is fairly good. Most CNS mastitis can be prevented through good milking procedures, accurate management measures and overall hygiene [1] . SCM caused by CNS can be locally treated with antibiotics during the period. However, of concern is the increasing prevalence of ß-lactamase-producing CNS [2] which are resistant to penicillins. Although SA is often sensitive to penicillin, such treatment may be unsuccessful due to the bacteria's ability to hide deep in the udder tissue, thus leading to chronic SCM [1] [4] .

Dairy cattle with chronic infection or with penicillin resistant bacteria should be eliminated to avoid new infections. Various antibiotics-based intra-mammary infusions for both dry-cow therapy and treatment of clinical mastitis are available on the Ugandan market. However, the increasing emergence of antibiotic resistant pathogens in the country [2] [7] [8] is further suspected to complicate the effectiveness of mastitis treatment. The treatment of clinical mastitis by farmers using antibacterial drugs without seeking prescription was reported [2] [7] as predisposing factor to the emergence of drug resistance and treatment failures. Therefore, this study established the prevalence of SCM and determined the potent antibacterial formulation for possible control of SA related SCM in selected dairy cattle farms in Kiboga district.

2. MATERIALS AND METHOD

2.1. Study Design

This was a basic cross-sectional study that was carried out between August and November 2012 to determine the prevalence, causes and choice of antibacterial drug for effective control of SA related SCM in Kiboga district. The study area included three randomly selected sub-counties of Kapeke, Lwamata, and Kiboga Town-Council.

2.2. Sample Size Determination

The sample size for this study was determined using the formula by Martinez et al., [9] at a previous prevalence of sub-clinical mastitis of 86.2% at dairy cattle level [5] . Furthermore, 5% Error at 95% confidence was considered for the calculation. The list of dairy farms in each sub-county was obtained from the District Veterinary Office and used as a sampling frame for random selection of farms for sampling. A total of 12 farms were chosen with 2 from Kapeke, 7 from the Town Council, and 3 from Lwamata. The selection criteria for these farms were based on herd structure, district veterinary records and accessibility as advised by the District Veterinary Office. All lactating dairy cattle in the selected herd that had no history of clinical mastitis were eligible for selection. However, lactating cattle that were on antibiotic therapy in the last one month prior to the study were excluded.

2.3. Selection of Cows and SCM Screening Using CMT

In each herd, the number of lactating dairy cattle that were sampled was determined by the herd size. A preliminary visit was made to the farms within the sub-counties of Kiboga district to ascertain the average herd size for small, medium and large herd categories in the area. Individual herds of dairy cattle were selected randomly and the entire herd was selected if it met the inclusion criteria. The cattle were restrained in a crush to avoid injury whenever found available on the farms. Each teat and its orifice were thoroughly cleaned with cotton wool soaked in 70% ethanol starting with teats that were furthest. Using a gloved hand, approximately 2 ml of milk from each quarter (starting with the furthest teats) were squirted in corresponding quadrants of the CMT paddle. An equivalent volume (2 ml) of CMT reagent was then added to each quadrant and mixed gently. The results were then interpreted qualitatively depending on the thickness of the gel formed [10].

2.4. Collection of Milk Samples for Microbial Analysis

All dairy cattle that were lactating were included in the study after analyzing their individual clinical history against the inclusion criteria. Milk samples were collected from all the quarters for bacterial culture and isolation. The teats and the orifices of the quarters were cleaned again using cotton wool soaked in 70% ethanol as described above. Approximately 2 ml of milk was stripped using clean gloved hands into sterile sample vial from the affected quarter(s) to make a composite sample. Each sample was labeled and the information regarding the cow was entered in the data sheet. The sample was immediately transferred into a cool box maintained at 5°C with ice packs and then taken to Kiboga District Veterinary Laboratory for bacterial analysis.

2.5. Bacterial Analysis

2.5.1. Microbial Culture of Milk Samples

Bacterial culture and isolation was done according to methods by Mornica [11] with minor modifications [1] . Briefly, 50 μl of composite milk was inoculated on 5% sheep blood agar medium and incubated for 18 - 20 hrs at 37°C. Identification of the bacteria was carried out basing on colony characteristics such as size, shape, color and hemolysis patterns [1] .

2.5.2. Antimicrobial Susceptibility Test for SA

This was done using Mueller-Hinton agar (MHA) according to Kirby-Bauer disc diffusion method [12] . The SA inoculum for each dairy farm was prepared as follows; one SA colony was mixed with 100 μl of distilled sterile water in sterile tubes. The mixture (inoculant) was applied on MHA plates and surface spread using a sterile spreader. Antibiotic discs were then applied using sterile forceps. The zones of inhibition were measured using a ruler and Oxford mathematical set divider, and the resultant diameters compared to the European Committee on Antimicrobial Susceptibility Testing [13] reference values for the different antimicrobial agents.

2.6. Questionnaire and Observations of Farm Practice

An open ended questionnaire was used to determine the hygiene status, occurrence of clinical mastitis as well as antibiotic use practices on the farm. The key lead questions were on milking practices, udder hygiene, personnel hygiene, knowledge on mastitis prevention and therapy. Observation of hygiene and milking practices were made for purposes of validation of the farmers' responses.

2.7. Data Analysis and Presentation

The data was analyzed descriptively using SPSS Version 16 software and presented as frequency tables.

3. RESULTS

3.1. Demographics

A total of 124 dairy cattle were screened for SCM from three sub-counties in Kiboga district. The majority (40%) of the dairy cattle sampled were from Lwamata while the rest were from Kapeke (30%) and Town Council (30%). Majority (70%) of the farms had over 20 heads of dairy cattle in both Lwamata and Kapeke. However, the Town Council was dominated by only small-holder dairy farmers who owned less than 10 heads of cattle. The overall proportion of lactating dairy cattle in the herd for the three sub-counties was 40%.

3.2. Prevalence of SCM

The prevalence of SCM for the three sub-counties ranged from 80% to 94.59%. Kapeke sub-county had the highest burden of SCM at a prevalence of 94.59% followed by Town Council with 91.89%; Lwamata recorded a prevalence of 80%. The district prevalence was found to be at 87.9% (Table 1).

3.3. Number of Udder Quarters Affected by SCM

Over 70% of the dairy cattle screened for SCM had more than 2 quarters affected. Majority (54.8%) of them had all the 4 quarters diagnosed with SCM while 21.0% of the cows had 3 quarters affected. However, 6.5% and 13.7% of the dairy cattle tested had SCM in 1 and 2 quarters of the udder respectively. Only 4% of all the dairy cattle screened tested negative for SCM (Table 2).

Table 1. Distribution of dairy cattle sampled for SCM across the study area.

Sub-county	No. of dairy cattle screened	No. of dairy cattle with SCM	% of dairy cattle SCM
Kapeke	37	35	94.59
Town council	37	34	91.89
Lwamata	50	40	80
Total	124	109	87.9%[a]

Key:[a] = Overall prevalence for the district; No. = Number.

Table 2. Number of udder quarters affected by SCM in the different dairy farms.

Location	Farm	1 Qtr	2 Qtrs	3 Qtrs	4 Qtrs	C.Qtrs	N
Kapeke	F1	0	1	3	13	0	17
	F2	0	3	8	9	0	20
	F3	0	0	0	3	0	3
	F4	0	0	1	4	0	5
	F5	1	4	1	2	0	8
Town Council	F6	0	2	1	1	0	4
	F7	0	0	1	3	0	4
	F8	0	0	1	3	0	4
	F12	1	1	2	4	1	9
	F9	4	1	7	16	2	30
Lwamata	F10	0	1	1	5	1	8
	F11	2	4	0	5	1	12
Total		8	17	26	68	5	124
Percentage		6.5	13.7	21.0	54.8	4.0	100

Key: F_1-F_{12} = farms; 1 Qtr-Dairy cattle in which one quarter of the udder had SCM, 2 Qtrs-Dairy cattle in which two quarters of the udder had SCM, 3 Qtrs-Dairy cattle in which three quarters of the udder had SCM, 4 Qtrs-Dairy cattle in which all the Four quarters had SCM and C. Qtrs-Dairy cattle in which all the 4 quarters had no SCM, N is the sample size from individual farms sampled.

3.4. Degree of Severity of SCM Based on CMT Scores

Over 60% of the cows tested had CMT scores ranging from 1 to 2 in the milk. However, 16.13% of the dairy cattle had more severe SCM as evidenced by scores of 3 in the milk as shown below (Table 3).

3.5. Bacterial Pathogens Isolated from Milk in the Different Dairy Farms Visited

The most prevalent bacterial pathogens associated with SCM were Staphylococci such as Coagulase negative Staphylococci (64.42%) and SA (16.56%). Streptococci species where below 10%; SAG at 3.68% and 5.52%

for the other Streptococci. However, other bacterial pathogens isolated included CB (6.13%), and CO (3.68%) as shown in the Table 4.

3.6. Susceptibility of SA against Common Antibacterial Drugs Used for Treatment of Mastitis

Majority (71.4%) of SA isolated were resistant to at least four antibacterial drugs. The prevalence of drug resistance was over 70% in the different farms in the Town Council compared to those in Kapeke and Lwamata in which resistance ranged from 42.6% to 57.1%. The SA isolated from all the 7 farms were highly resistant to Penicillin (100%) followed by Neomycin (85.7%) and tetracycline (71.4%). Resistance to Streptomycin and Gentamicin were 42.6% and 28.6% respectively. Trimethoprim-Sulphamethoxazole (Sulfamethoxazole), was potent and no resistance was recorded in SA isolated from the 7 farms in Kiboga district (Table 5).

3.7. Hygiene Status, Milking Practices and Management of Mastitis on the Dairy Farms

The factors that could have contributed to high prevalence of SCM included; lack of knowledge on SCM (100%), cleaning towels (100%), teat dipping (100%), dry cow therapy (100%) and teat pulling (100%) during milking as well as no usage of disinfectant (100%). Over 50% of the farmers had poor farm hygiene and visible dirt on the udder of lactating cows. Antimastitic formulations containing penicillins (60%), tetracycline (30%) sulfonamides (5%) and aminoglycoside (5%) were used for treatment of clinical mastitis. Unfortunately, 68% of the treatment was done by the farmers and herdsmen while only 32% was done by veterinarians (Table 6).

Table 3. Severity of subclinical mastitis in the various farms based on CMT scores.

		CMT score				
Location	Farm	0	1	2	3	N
Kapeke	F_1	7	26	12	23	68
	F_2	16	27	22	15	80
Town council	F_3	0	4	4	4	12
	F_4	1	8	6	5	20
	F_5	13	11	3	5	32
	F_6	5	9	2	0	16
	F_7	1	8	6	1	16
	F_8	1	8	3	4	16
	F_{12}	10	11	9	6	36
Lwamata	F_9	29	46	35	10	120
	F_{10}	7	17	5	3	32
	F_{11}	17	17	10	4	48
Total		107	192	117	80	496
Percentage		21.57	38.71	23.59	16.13	100

Key: F_1-F_{12}-farms, 0-negative, no thickening, 1-distinct thickening, no gel formation; 2-distinct thickening, slight gel formation; 3-gel is formed; N is the sample size.

Table 4. Prevalence of bacteria isolated from milk in the different dairy farms.

Farm (F)	SA	CNS	SAG	OS	CB	CO
F_1	5	12	2	3	0	3
F_2	11	18	3	2	0	1
F_3	0	3	0	0	0	1
F_4	2	5	0	1	0	0
F_5	1	8	0	0	0	0
F_6	0	4	0	0	0	0
F_7	1	4	0	0	0	0
F_8	0	4	0	0	0	0
F_9	6	20	0	2	10	1
F_{10}	1	8	1	1	0	0
F_{11}	0	12	0	0	0	0
F_{12}	0	7	0	0	0	0
Total:	27	105	6	9	10	6
Prev.	16.56	64.42	3.68	5.52	6.13	3.68

Key: F_1-F_{12}-farms; SA = *Staphylococcus aureus*, CNS = Coagulase negative staphylococcus; SAG = *Streptococcus agalactiae*, OS = Other Streptococcus, CB = *Corynebacteriumbovis;* CO = Coliforms; Prev. = prevalence; total number of bacterial isolates = 163.

Table 5. Susceptibility of Staphylococcus aureus against common antibacterial drugs used for treatment of mastitis in Uganda

Location	Farm	PG10C	T30C	G10C	CX5C	SXT25	N10C	S10C
Kapeke	F_1	R	R	S	R	S	S	S
	F_2	R	R	S	R	S	R	S
Town Council	F_4	R	R	R	R	S	R	R
	F_5	R	R	S	R	S	R	R
	F_7	R	R	R	R	S	R	S
Lwamata	F_9	R	S	S	R	S	R	S
	F_{10}	R	S	S	R	S	R	R
	%RD	100	71.4	28.6	100	0	85.7	42.6

Key: F_1-F_{10} = farm; R = resistant, S = susceptible: PG10C = penicillin G; T30C = tetracycline: G10C = gentamicin; CX5C = cloxacillin: SXT25 = trimethoprimsulphamethazole: N10C = neomycin; S10C = streptomycin; % RD = Resistance per drug basis.

4. DISCUSSION

The current prevalence of SCM in Kiboga district (87.9%) was higher than what has been previously reported. Byarugaba et al. [6] found a prevalence of 61.3% in Jinja district. In a related study, Abrahmsén [5] reported prevalence of SCM of 86.2% among dairy cattle reared in urban and peri-urban areas of Kampala. Byarugaba et al. [6] , carried out the study on mastitis on dairy farms that were predominantly managed under zero grazing system unlike the current study where cattle were kept under extensive system. Within Kiboga district, variations in prevalence of SCM among different sub-counties could be due to the extent of accessibility to extension services and clean water. Highly contaminated water carries bacteria that can gain entry into the udder to cause SCM when used for cleaning the teats especially with the absence of post teat dipping.

The study also revealed that 70% of the dairy cattle tested had SCM in more than 2 quarters of the udder and 54.8% having all the 4 quarters affected. This has a serious implication on economics of milk production in the affected dairy farms. It is recognized that SCM can cause reduction in milk production from the affected quarter by up to 60% [1] . Moreover, at least 60% of the dairy cattle tested had CMT scores ranging from 1 to 2. This indicates that majority of the dairy cattle that tested positive had severe SCM. Similarly, Hand et al. [14] reported significant milk production losses associated with increase in SCC due to SCM. Other economic losses attributed to SCM include increase in treatment cost and culling [15] .

Table 6. Hygiene status, milking practices and management of mastitis on the farms.

Variable		Frequency of response by dairy farmers			
		Scale/action	Kapeke	Town Council	Rwamatta
		Good	0	2	1
	Cleanliness of resting area	Fair	0	2	0
		Poor	2	3	2
	Personal hygiene	Good	1	4	1
		Fair	1	0	0
Hygiene status		Poor	0	3	2
Udder hygiene	Dirty		1	4	2
	Relatively clean		1	3	1
Milking practice	Use of cleaning towel		0	0	0
		Fore-striping	2	2	2
		Teat pulling	2	7	3
		Calf stimulation	2	7	3
		Hand milking	0	0	0
	Disinfection of milking equipment		0	0	0
Mastitis prevention	Teat dipping		0	0	0
	Dry cow therapy		0	0	0
Drug types	Multiject®		2	7	3
	Duofast®		1	0	0
	Gentamast®		0	0	1
	Tetracycline		2	2	2
Who treats mastitic cows	Herdsman		2	4	0
	Area veterinarian		2	4	0
	Farmer		0	4	3
What happens to the milk during drug withdrawal period	Home use		2	4	0
	Feed to the dogs		0	1	3
	Sold		0	2	0
	Poured away		0	1	1

In this study, the major constraints to mastitis control were primarily farmers' poor hygienic practices, largely attributed to lack of knowledge on the disease. No wonder, all the farmers assessed had no prior knowledge about SCM although they were aware of clinical mastitis. Poor environment and udder hygiene characterized by extensive soiling of the udder and the teats with manure were amongst factors predisposing cattle to SCM. This was further exacerbated by poor personnel hygiene; contaminated water coupled with lack of cleaning towels thus compounding the spread of contagious and environmental pathogens that cause SCM. Moreover, the few dairy farms that used milking salve had high level of bacterial load during preparation due to poor hygiene. Unfortunately, such poor hygiene also leads to possible contamination of milk thus arousing genuine milk safety concern.

While Uganda Dairy Development Authority (DDA), urges all dairy producers to use metallic buckets, a majority (10/12) of the farms assessed use plastic milking jars which are difficult to effectively clean thus encouraging bacterial multiplication. Bacterial culture results indicate that the most prevalent causative pathogens associated with SCM were Staphylococci (over 80%). Anri [1] also reported that Staphylococci are the dominant causes of SCM and are associated with low CMT scores. The contagious pathogens isolated were SA (16.56%), SAG (3.68%), CB (6.13%) which accounts for the high prevalence of SCM in the dairy cattle and being propagated by the poor hygienic practices at the dairy farms visited. However, environmental pathogens isolated included: CNS (64.42%), OS (5.52%), and CO (3.68%). Since CO are usually associated with clinical mastitis, this proportion was probably due to the contamination of the milk samples during sample collection [1] .

Antibacterial sensitivity of SA isolated from the 7 dairy farms revealed 100% resistance against penicillin followed by neomycin (85.7%) and tetracycline (71.4%). At least 71.4% of SA isolated from the above farms were resistant to at least four antibacterial drugs. This multi-drug resistance is a real threat to effective control of the disease since the options are limited. Such high prevalence of resistance by udder pathogens was previously reported [6] . The highest level of resistance was registered from samples obtained from the Town Council (over 50%), suggestive of high drug pressure and irrational use of antibiotics. Indeed, over 80% of the farmers used Multiject□(penicillin, neomycin and streptomycin) and tetracycline for treatment of mammary and systemic bacterial diseases. The sale of antibiotics over-the-counter without prescription [16] has offered unprecedented access to the drugs by farmers. No wonder, only 30% of the farmers relied on veterinarians for treatment but the majority treated their animals with the aid of herdsmen. Such treatments were carried without strict adherence to dosage requirements and treatment period due to lack of technical knowledge among farmers and herdsmen considering their low level of education. Such abuse of drugs is responsible for the emergence of multi-drug resistant SA [17] which is a threat to both animal and public health. However, Trimethoprim-sulphamethazole was 100% effective against penicillin, tetracycline and aminoglycoside resistant SA.

5. CONCLUSION

The high burden of SCM and emergence of multidrug resistant SA are one of the constraints to dairy production in Kiboga district. Therefore, sensitization of dairy farmers in Kiboga district on proper hygiene, appropriate milking techniques and dry cow therapy based on the results of antibiotic sensitivity tests are highly recommended in SA associated SCM. Further research is also needed to identify the most appropriate extension materials on mastitis that can have a significant impact on effective control of the disease.

ACKNOWLEDGEMENTS

The authors would like to acknowledge the contribution of Japan International Cooperation Agency (JICA) through the Animal Disease Control project towards the success of this study.

REFERENCES

1. Anri, A. (2012) Manual for Mastitis Control in Developing Countries. Japan Livestock Technology Association. 37-56.

2. Byarugaba, D.K., Kisame, R. and Olet, S. (2011) Multi-Drug Resistance in Commensal Bacteria of Food of Animal Origin in Uganda. African Journal of Microbiology Research, 5, 1539-1548.

3. Vaarst, M., Byarugaba, D.K., Nakavuma, J. and Laker, C. (2007) Participatory Livestock Farmer Training for Improvement of Animal Health in Rural and Peri-Urban Smallholder Dairy Herds in Jinja, Uganda. Tropical Animal Health and Production, 39, 1-11.

4. Pyörälä, S. (1995) Staphylococcal and Streptococcal Mastitis. In: Sandholm, M., Honkanen-Buzalski, T., Kaartinen, L., Pyörälä, S., Eds., The Bovine Udder and Mastitis, Faculty of Veterinary Medicine, University of Helsinki, Helsinki, 143-148.

5. Abrahmsén, M. (2012) Prevalence of Subclinical Mastitis in Dairy Farms in Urban and Peri-Urban Areas of Kampala, Uganda. Second Cycle. SLU, Department of Clinical Sciences, Uppsala. http://stud.epsilon.slu.se/4213/1/Abrahms%C3%A9n_m_120507.pdf

6. Byarugaba, D.K., Nakavuma, J.L., Vaarst, M. and Laker, C. (2008) Mastitis Occurrence and Constraints to Mastitis Control in Smallholder Dairy Farming Systems in Uganda. Livestock Research for Rural Development, 20, 1.http://www.lrrd.org/lrrd20/1/byar20005.htm

7. Nakavuma, J., Byarugaba, D.K., Musisi, L.N. and Kitimbo, F.X. (1994) Microbiological Diagnosis and Drug Resistance Patterns of Infectious Causes of Mastitis. Uganda Journal of Agricultural Sciences, 2, 22-28.

8. Kintu, A., Byarugaba-Karuhize, D. and Nalwanga, D.K. (2000) Mastitis in Peri-Urban Zero-Grazing Dairy Farming around Kampala. Uganda Veterinary Journal, 6, 45-50.

9. Martinez, G.R. (1988) An Appraisal of Mastitis and the Potential for Its Control in Dairy Herds on the Savannah of Bogota. PhD Thesis, University of Reading, Reading.

10. Mellenberger, R. and Roth, C.J. (2000) California Mastitis Test (CMT), Fact Sheet. Department of Animal Sciences, Michigan State University and Department of Dairy Science, University of Wisconsin-Madison.

11. Monica, C. (2006) District Laboratory Practice in Tropical Countries, Part 2. 2nd Edition, 62-70, 132-143, 157-234.

12. Quinn, P.J., Carter, M.E., Markey, B. and Carter, G.R. (1994) Clinical Veterinary Microbiology. Wolfe Publishing, London.

13. EUCAST (2012) Disk Diffusion Method for Antimicrobial Susceptibility Testing. Version 2.1. www.eucast.org

14. Hand, K.J., Godkin, A. and Kelton, D.F. (2012) Milk Production and Somatic Cell Counts: A Cow-Level Analysis. Journal of Dairy Science, 95, 1358-1362.http://www.sciencedirect.com/science/ article/pii/S0022030212000926

15. Jemeljanovs, A., Cerina, S. and Konosonoka, I.H. (2010) Economic Losses Resulting from Subclinical Mastitis in Dairy Cows.http://www.eaap.org/previous_annual_meetings/2010crete/papers/18_jemeljanovs.pdf

16. Byarugaba, D.K. (2004) A View on Antimicrobial Resistance in Developing Countries and Responsible Risk Factors. International Journal of Antimicrobial Agents, 24, 105-110.

Index